CHEMISTRY RESEARCH AND AP

EUROPIUM III: DIFFERENT EMISSION SPECTRA IN DIFFERENT MATRICES, THE SAME ELEMENT

CHEMISTRY RESEARCH AND APPLICATIONS

Additional books in this series can be found on Nova's website under the Series Tab.

Additional E-books in this series can be found on Nova's website under the E-books tab.

CHEMISTRY RESEARCH AND APPLICATIONS

EUROPIUM III: DIFFERENT EMISSION SPECTRA IN DIFFERENT MATRICES, THE SAME ELEMENT

EDUARDO J. NASSAR
KATIA J. CIUFFI
AND
PAULO S. CALEFI

Novinka
Nova Science Publishers, Inc.
New York

NOTICE TO THE READER

LIBRARY OF CONGRESS CATALOGING-IN-PUBLICATION DATA

Available upon Request
ISBN: 978-1-61728-306-2

Published by Nova Science Publishers, Inc. ✝ New York

Contents

PREFACE

Europium belongs to the lanthanide series of the periodic table. The most important property of this element is its luminescence, which can be defined as emission of light by a material as a consequence of energy absorption. Such property leads to the main use of this element; that is, its application as phosphor, which emits colored light through electroluminescent processes, with potential utilization in computer and TV displays. Many scientists employ the luminescence spectroscopy of the europium ion to elucidate the structure of natural and synthetic compounds. In this chapter, we propose a review on the photoluminescence of the europium ion in different materials, prepared by hydrolytic and non-hydrolytic sol-gel methodologies and intercalation in kaolinite. The sol-gel route is well-known for its simplicity and high rates. It is the most commonly employed technique for the synthesis of nanoparticles, and it involves the simultaneous hydrolysis and condensation reaction of the alkoxide or salt. The obtained materials have several particular features. The importance and advantages of nanoparticles have been scientifically demonstrated, and these particles have several industrial applications; e.g. in catalysts, pigments, biomaterials, phosphors, photonic devices, pharmaceuticals, among others. In this chapter, the sol-gel route was utilized to prepare hybrid materials, biomaterials, thin films, phosphors, and intercalation materials. Throughout the entire chapter, either the europium ion or europium compounds were used in the preparation of luminescent materials or structural probes.

Chapter 1

INTRODUCTION

In the nineteenth century, a new group of elements was discovered, but only in the twentieth century were all rare earths known and their special properties explored [1]. In 1901, Demarcay discovered the europium element [2], but he did not have the slightest idea of the counteless applications of this element in its oxidation forms Eu^{2+} and Eu^{3+}. Europium belongs to the lanthanide series of the periodic table. The most important property of this element is its luminescence, derived from the incompletely filled 4f shell electrons. These electrons are shielded by the 5s and 5p closed shells, so they do not participate directly in bonding and interact much less strongly with the environment. Europium ions have been explored due to their photoluminescent properties, which result in the emission of sharp atomic bands corresponding to $f \rightarrow f$ transitions in the central metal ion [3]. Basic to the understanding of Eu^{3+} ion luminescence experiments is the appropriate energy level diagram depicted in Figure 1.

The emission lines for the Eu^{3+} ion usually appear in the visible region. These lines correspond to transitions from the excited $^5D_0$ level to the ^7F_J (J = 0, 1, 2, 3, 4) levels of the $4f^n$ configuration. The red spectral area 610 nm ($^5D_0 \rightarrow {}^7F_2$ transition) is the main emission of the Eu^{3+} ion. The large magnitude of spin-orbit coupling in europium causes the individual J levels of the electronic terms to be well separated from each other [4].

Compared with other lanthanide ions, the $^5D_0 \rightarrow {}^7F_2$ hypersensitive transition in Eu^{3+} is the most sensitive to the environment. This phenomenon can be explained by possible mixing of the electron transfer character into the hypersensitive transition. The interaction between the Eu^{3+} ion with its surrounding usually induces perturbation or modification in luminescence parameters such as intensity, quantum efficiency, and lifetime of the ion emission. These perturbations in luminescence allow for the explanation and understanding of the nature and origin of the interactions between the Eu^{3+} ion and the different surroundings. The intensity, position of the excitation and emission wavelengths, and lifetime of the transition are some variables that characterize the system, since these properties are intrinsic and can be modified according to the environment present around the Eu^{3+} ion.

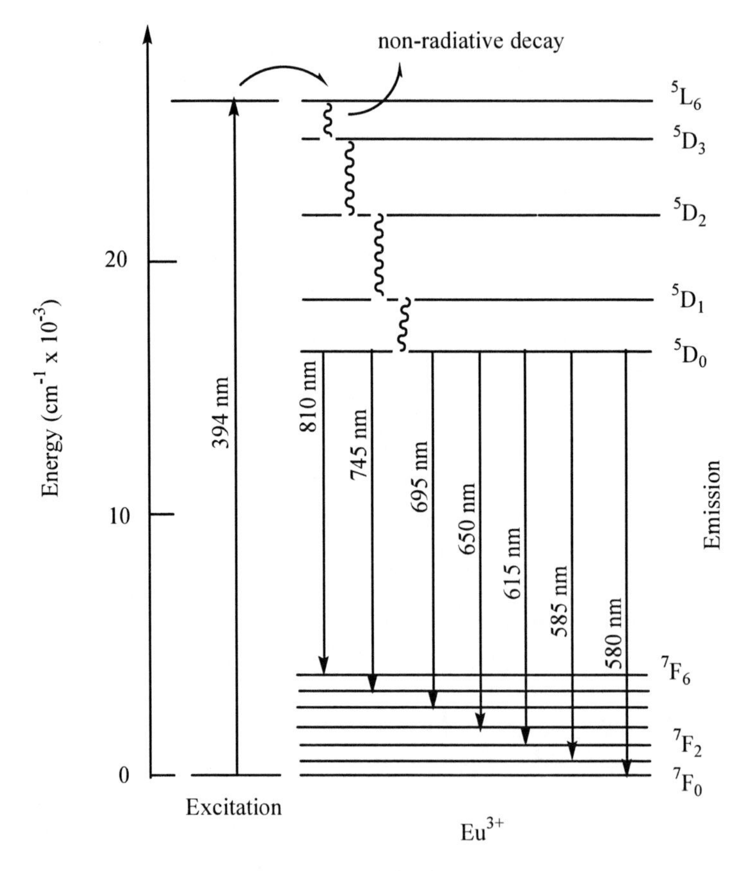

Figure 1. Energy Level Diagrams For The Eu^{3+} Ion.

The sol-gel technology is employed in the preparation of inorganic ceramic and glass materials. Its beginning dates back to the mid 1800s with Ebelman and Graham´s studies on silica gels [5]. Initially, the sol-gel process studied the preparation of silicate from tetraethylorthosilicate (TEOS, $Si(OC_2H_5)_4$), which is mixed with water and a mutual solvent, to form a homogeneous solution. Recently, new reagents such as titanium, aluminum, vanadium, yttrium, and other alkoxides have appeared, so novel inorganic oxides can be prepared using this methodology. Another process known as non-hydrolytic sol-gel was developed by Acosta et al [6], who used the condensation reaction between a metallic or semi-metallic halide (M-X) and a metallic or semi-metallic alkoxide (M'-OR) to obtain an oxide (M-O-M'). The alkoxide can be added to the reaction or produced *in situ* during the process through reaction between the halide and an oxygen donor, such as alcohol or ether. The hydrolytic and non-hydrolytic sol-gel processes as well as their mechanisms are well discussed in the literature [5, 7 - 9].

The sol-gel route is well-known for its simplicity and high rates. It is the most commonly employed technique for the synthesis of nanoparticles, and it involves the simultaneous hydrolysis and condensation reaction of the alkoxide or salt. The obtained materials have several particular features. The importance and advantages of nanoparticles have been scientifically demonstrated, and these particles have several industrial applications; e.g. in catalysts, pigments, biomaterials, phosphors, photonic devices, pharmaceuticals, among others [9].

In this chapter, we propose a review on the photoluminescence of the Eu^{3+} ion in different materials, prepared by hydrolytic and non-hydrolytic sol-gel methodologies and intercalation in kaolinite. We will show all the research developed in our laboratory using the sol-gel route to prepare hybrid materials, biomaterials, thin films, phosphors, and intercalation materials. Throughout the entire chapter, either the europium ion or europium compounds were used in the preparation of luminescent materials or structural probes.

Chapter 2

Hydrolytic Sol-Gel Process

The hydrolytic sol-gel process is utilized to obtain materials with optical applications. This route is based on the inorganic polymerization of molecular precursors and involves the evolution of a polymer through a colloidal (sol) suspension with the formation of a net, going through the continuous liquid phase (gel). The molecular precursor used in the sol-gel process is essentially an $M(OR)_z$ alkoxide, where M is silicon or a metal and R is an alkyl group. Several combinations can be obtained depending on the employed alkoxide, and hybrid organic-inorganic materials containing different combinations of alkoxides and organo-alkoxides can be achieved by using a combination of tetraethylorthosilicate (TEOS) with such silanes as phenyltriethoxysilane (PTES) [10, 11], 3-methacryloxypropyltrimethoxysilane (MPTMS) [12, 13], and 3-aminopropyltriethoxysilane (APTS) [14].

Materials containing inorganic components, such as calcium and phosphate ions, can be used as biomaterials [15, 16]. Nanoparticles can also be prepared by the hydrolytic sol-gel methodology [17]. Other applications of this process are the preparation of coatings or thin films, due to its facility when working with solutions [18 - 21].

Non-Hydrolytic Sol-Gel Process

The basis of the nonhydrolytic sol-gel method is the condensation reaction between a metallic or semi-metallic halide (M-X) and a metallic or semi-metallic alkoxide (M'-OR), which leads to the formation of an oxide (M-O-M'), the by product of this reaction being an alkyl halide (R-X). This process allows production of several kinds of oxides with different compositions, such as glass ionomers used in odontology, at low temperature [22, 23]. The low temperature employed in the process and the mixture at molecular level are advantages of this methodology, which can be utilized in the preparation of phosphors [24 - 30].

In all the syntheses, the ethanolic europium chloride solution ($EuCl_3$) was added to the sol in molar percentages of 1.0 %, and the sol was homogenized by magnetic stirring.

Chapter 3

CHARACTERIZATION TECHNIQUES

All the samples were characterized by the different techniques described below:

Thermogravimetric analyses (TG/DTA/DSC) were accomplished in a Thermal Analyst 2100- TA Instruments SDT 2960 with Simultaneous TG/DTA/DSC, in nitrogen atmosphere, at a heating rate of 20°C/min, from 25 to 1000 °C.

X-ray diffractograms (XRD) were obtained at room temperature in a Siemens (D 5005) or a Rigaku Geigerflex D/max-c diffractometer system using monochromated CuK_{α} radiation (λ=1.54 Å) over the 2θ range between 4 and 80°, at a resolution of 0.05°.

^{27}Al, ^{29}Si, and ^{13}C NMR (59.5 MHz) analyses were carried out in an INOVA 300 Varian spectrophotometer, using silicon nitride as reference.

The morphology of the systems was investigated by transmission electron microscopy (TEM). To this end, a drop of powder suspension was deposited on a copper grid. TEM analysis was performed using a 200 kV Philips CM 200 microscope. Scanning electron microscopy (SEM) images and energy dispersive X-ray spectrometry (EDS) were recorded on a Hitachi S-4100 microscope.

The infrared absorption spectra were obtained with a Perkin-Elmer 1739 spectrophotometer with Fourier transform, using the KBr pellet technique.

The refractive index and thickness of the waveguides were measured by both transverse electric (TE) and transverse magnetic (TM) polarization with an m-line apparatus (Metricon Model 2010) based on the prism coupling technique.

The electronic spectra of the films were recorded in an ultraviolet visible spectrophotometer (Hewlett-Packard 8453, diode array). The spectra were obtained directly, using the substrate as blank.

The symmetry site was studied by photoluminescence data. Excitation and emission spectra at room temperature were obtained under continuous (450 W) Xe lamp excitation in a SPEX Fluorolog Spectrofluorometer. All spectra were corrected by spectrometer optics, lamp output, and detector response. Decay curves were processed with accessory SPEX 1934 phosphorimeter equipped with pulsed (5 J/pulse) Xe lamp or in a Jobin Yvon-Spex spectrometer (HR 460) coupled to a R928 Hamamatsu photomultiplier. A Xe arc lamp (150 mW) coupled to a Jobin Yvon-Spex monochromator (TRIAX 180) was used as the excitation source. All the spectra were obtained at room temperature.

We will firstly discuss the influence of different organosilanes in the excitation and emission spectra of the Eu^{3+} ion in hybrid materials obtained by the hydrolytic sol-gel process. The influence of temperature, pre-hydrolysis, and other parameters will be considered. Four different silicon alkoxides were employed; namely tetraethylorthosilicate (TEOS), phenyltriethoxysilane (PTES), 3-methacryloxypropyltrimethoxysilane (MPTMS), and 3-aminopropyltriethoxysilane (APTS). The molecular structures of these silanes are shown in Figure 2.

The initial samples were prepared using the tetraethylorthosilicate (TEOS) and phenyltriethoxysilane (PTES) alkoxides. The first sol was prepared using a 1:1 TEOS/PTES molar ratio, while the second was prepared with PTES only. Ammonium hydroxide was used as catalyst. The obtained solids were washed and dried at 50°C for 24 hours. The samples were heat-treated for 4 hours at 100, 200, or 300°C.

Figure 3 illustrates the excitation spectra of the Eu^{3+} ion doped into silica matrices prepared with PTES or TEOS+PTES, monitored at 612 nm ($^5D_0 \rightarrow {}^7F_2$).

The lines ascribed to the transition corresponding to the manifold level $^7F_0$ to $^5L_6$ and $^5D_{1,\ 2\ \text{and}\ 3}$ excited level were observed. For the samples prepared with the PTES alkoxide only, there was a band in the high-energy region with maximum at 276 nm, which was assigned as charge transfer band (CTB).

There was a change in the excitation spectra of all the samples after a 4-hour heat-treatment at 100, 200, or 300°C. Figures 4 and 5 present the excitation spectra of the Eu^{3+} ion in the samples prepared with PTES and TEOS+PTES, respectively, heat-treated at different temperatures.

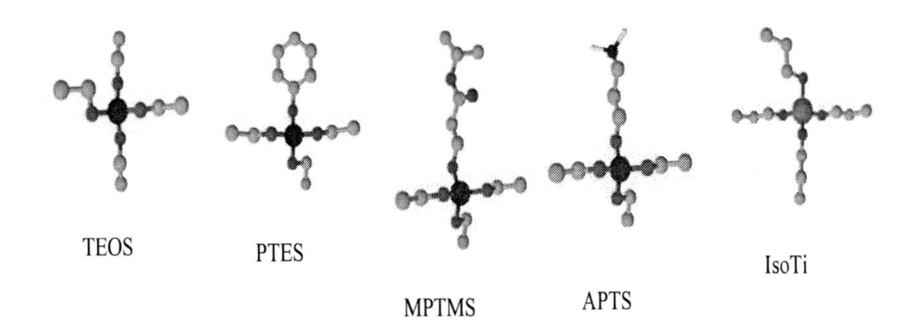

Figure 2. Molecular structure of the silicon alkoxides employed in this work.

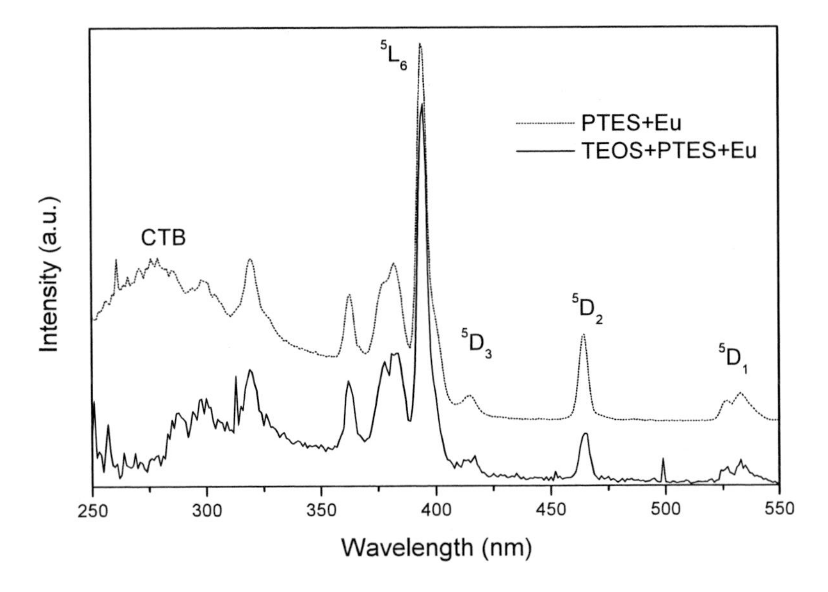

Figure 3. Excitation spectra at room temperature of Eu^{3+} ions doped into different matrices, $\lambda_{em} = 612$ nm.

The excitation spectra of the Eu^{3+} ion reveal the effect of the heat-treatment temperature. The broad band in Figure 4, ascribed to the CTB, appears at different wavelengths for the three samples. This band shifts to higher wavelength with increasing heat-treatment temperature, ongoing from a smaller band at 250 nm to 300 and 350 nm for the samples treated at 100, 200, and 300°C, respectively. These changes in the excitation spectra might be due to differences in the Eu^{3+} surrounding. The $^7F_0 \rightarrow {}^5L_6$ transition appears in the spectra of all the samples. Indeed, in the first article using Eu^{3+} ion as probe to investigate symmetry changes in the sol-gel process, Levy et al [31] reported the dependence of Eu^{3+} emission on heat-treatment temperature.

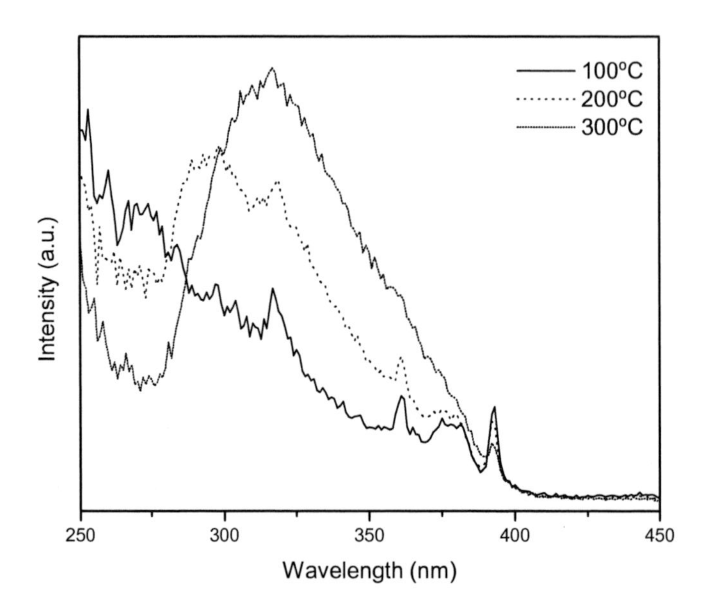

Figure 4. Excitation Spectra At Room Temperature Of The Eu^{3+} Ion In The Samples Prepared With PTES Alkoxide Only And Heat-Treated At 100, 200, Or 300°c, λ_{Em} = 612 Nm.

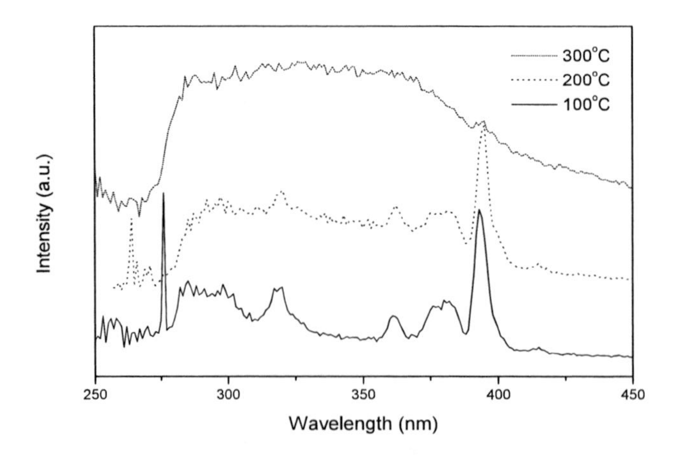

Figure 5. Excitation spectra at room temperature of the Eu^{3+} ion in the samples prepared with TEOS+PTES alkoxide and heat-treated at 100, 200, ord 300°C, λ_{em} = 612 nm.

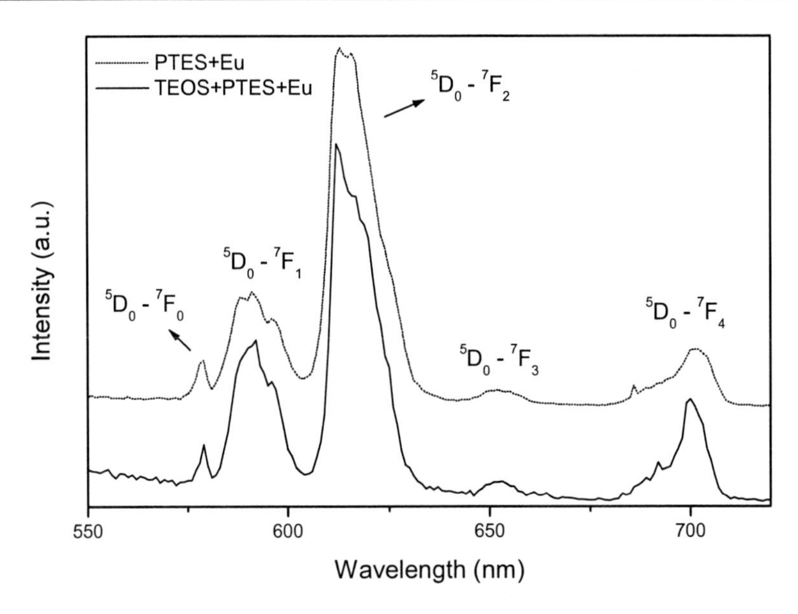

Figure 6. Emission spectra at room temperature of the Eu^{3+} ion in the samples prepared with PTES and TEOS+PTES alkoxide, λ_{exc} = 394 nm.

Figure 5 gives evidence of a more pronounced difference in the emission spectrum of Eu^{3+} in the TEOS+PTES sample heat-treated at a temperature of 300°C. At this temperature, the silica matrix can present several relaxation mechanisms due to decomposition of the organic component. The ^{29}Si NMR obtained for the sample treated at 300°C does not display the signal corresponding to the Si-C bond, which is one more indication of the onset of phenyl decomposition [10].

Figure 6 corresponds to the emission spectra of the Eu^{3+} ion in the samples prepared with PTES and TEOS+PTES alkoxide, monitored at 394 nm ($^5L_6$ level). Figures 7 and 8 represent the emission spectra of the samples after being heat-treated at 100, 200, or 300°C, for PTES and TEOS+PTES, respectively.

The emission spectra obtained by excitation at 394 nm ($^5L_6$ level) consist of the $^5D_0 \rightarrow {}^7F_J$ (J = 0, 1, 2, 3 and 4) emission lines of Eu^{3+} dominated by the $^5D_0 \rightarrow {}^7F_2$ (~ 610 nm) electric dipole transition, which is strongly dependent on the Eu^{3+} surrounding. When the Eu^{3+} ion is situated in a low symmetry site (without inversion center), the hypersensitive transition $^5D_0 \rightarrow {}^7F_2$ is often dominating in the emission spectrum. This indicates that the Eu^{3+} ion occupies a site without inversion center. The $^5D_0 \rightarrow {}^7F_1$ transition (591 nm) is purely magnetic-dipole allowed and not restricted by any symmetry [32, 33]. The Eu^{3+} emission bands in these spectra give evidence of a nonhomogenous distribution of the ion in the silica matrix [34 – 36].

The absence of TEOS influences the surrounding of the Eu^{3+} ion. In this situation, incomplete condensation is promoted due to the larger number of –OH groups on the silica surface. These –OH groups promote nonradiative energy loss by vibration modes.

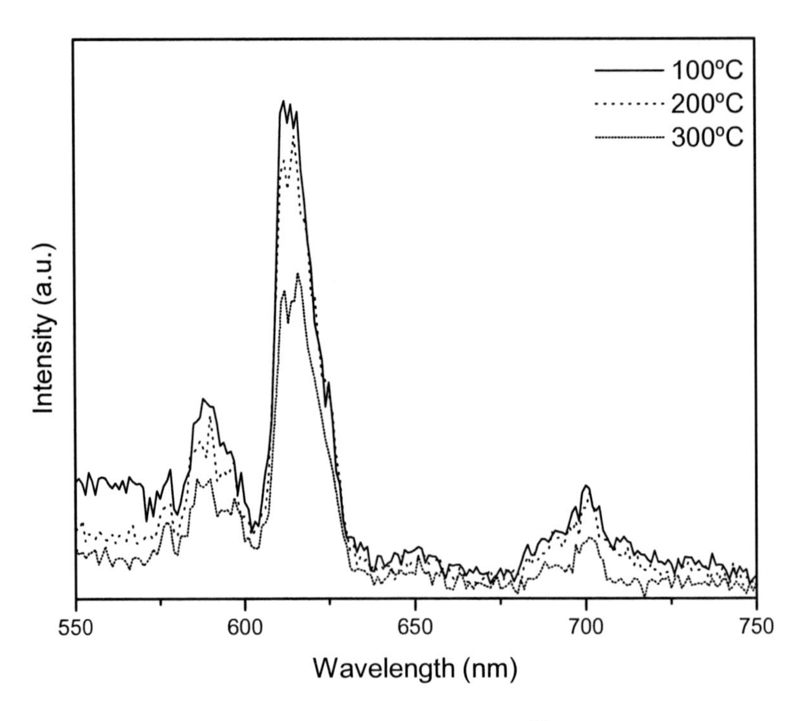

Figure 7. Emission spectra at room temperature of the Eu^{3+} ion in the samples prepared with PTES, heat-treated at 100, 200, or 300°C, $\lambda_{exc} = 394$ nm.

Because of the electric dipole character of Eu^{3+}, the intensities of the $^5D_0 \rightarrow {}^7F_0$ and $^5D_0 \rightarrow {}^7F_2$ transitions are strongly dependent on the ion's surrounding. The band corresponding to the $^5D_0 \rightarrow {}^7F_1$ transition, in turn, presents a magnetic dipole nature, so its intensity is not affected by the environment around Eu^{3+}. Therefore, the latter band can be considered a standard to measure the relative intensity of the other bands [37, 38]. Table 1 shows the relative intensity of the $^5D_0 \rightarrow {}^7F_0$ and $^5D_0 \rightarrow {}^7F_2$ transitions with respect to the $^5D_0 \rightarrow {}^7F_1$ transition, for all the samples.

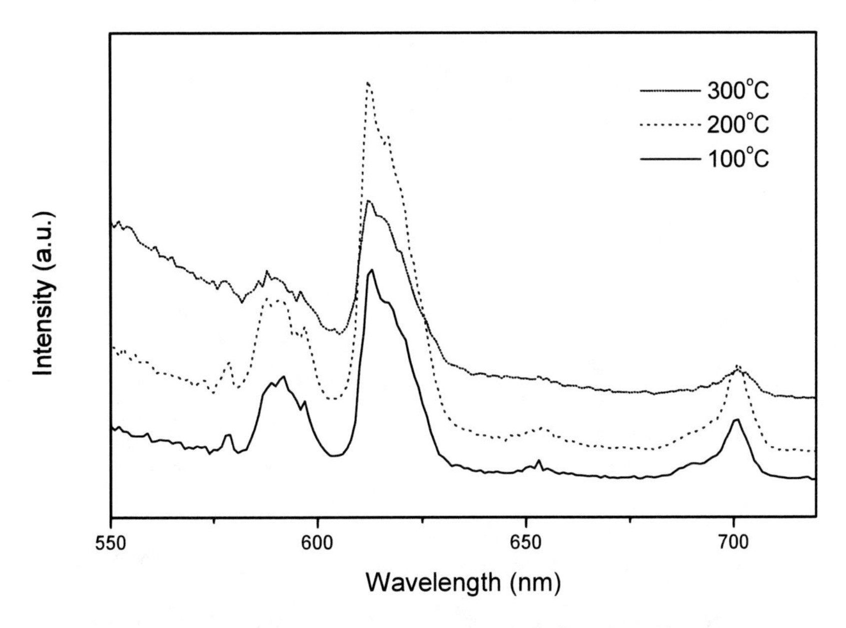

Figure 8. Emission spectra at room temperature of the Eu^{3+} ion in the samples prepared with TEOS+PTES, heat-treated at 100, 200, or 300°C, λ_{exc} = 394 nm.

The relative intensities of the bands for the samples containing no TEOS are similar, regardless of the heat-treatment temperature. This is because the Eu^{3+} surrounding is the same in all the three samples. However, the heat-treatment temperature does influence the spectrum of the samples containing TEOS. The decrease in the relative intensity $^5D_0 \rightarrow {}^7F_2/{}^5D_0 \rightarrow {}^7F_1$ indicates an increase in the symmetry around the Eu^{3+} ion. The scheme 1 can be represents the hybrid structure.

Table 1. Relative intensity ratio of the $^5D_0 \rightarrow {}^7F_0/{}^5D_0 \rightarrow {}^7F_1$ and $^5D_0 \rightarrow {}^7F_2/{}^5D_0 \rightarrow {}^7F_1$ transitions

Samples	$^5D_0 \rightarrow {}^7F_0/{}^5D_0 \rightarrow {}^7F_1$	$^5D_0 \rightarrow {}^7F_2/{}^5D_0 \rightarrow {}^7F_1$
TEOS+PTES 100°C	0.15	1.94
TEOS+PTES 200°C	0.28	1.55
TEOS+PTES 300°C	0.37	1.25
PTES 100°C	0.25	2.03
PTES 200°C	0.25	2.12
PTES 300°C	0.26	2.08

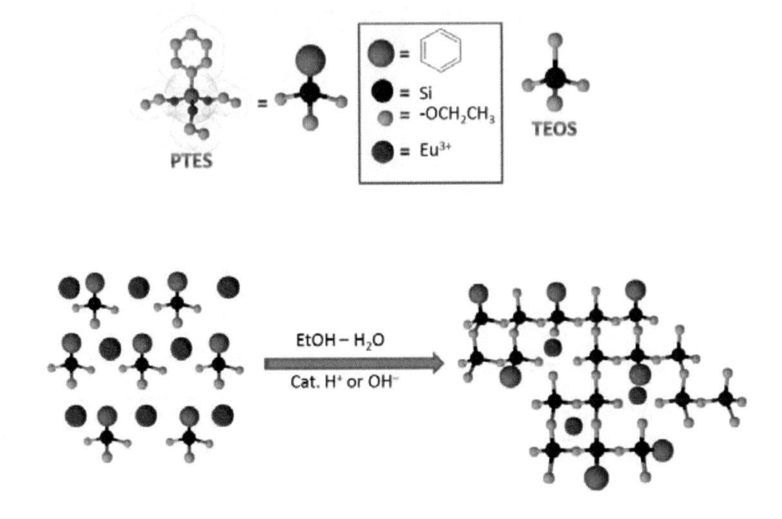

Scheme 1. Schematic representation of the hybrid sctructure contends PTES alkoxide.

The effect of phosphate on Eu^{3+} emission was studied in the hybrid materials prepared with TEOS, PTES alkoxide, and ammonium phosphate ($NH_4H_2PO_4$) at the following molar ratios: 1:0:0.25, 0:1:0.25, and 1:1:0.25; ammonium hydroxide was employed as catalyst. The solvent was evaporated at room temperature, monoliths were obtained, and the samples were dried at 50°C and triturated, resulting in xerogels (powders).

Figures 9 and 10 depict the excitation and emission spectra of the Eu^{3+} ion doped into silica containing different phosphate ion molar ratios. The photoluminescence behavior of the Eu^{3+} ions in the different samples was similar. The maximum at 393 nm in the excitation spectra of all the samples was ascribed to the $^5L_6$ level of Eu^{3+}. The emission spectra presented transitions arising from $^5D_0$ to the ^7F_J (J = 0, 1, 2, 3 and 4) manifolds excited at their maximum in all the samples.

The luminescence spectra of Eu^{3+} ions doped into xerogels containing phosphate ions presented similar luminescence bands, revealing the presence of wide bands and intensification of the $^5D_0 \rightarrow {}^7F_1$ transition compared with the samples prepared in the absence of phosphate ions. This enhancement in the $^5D_0 \rightarrow {}^7F_1$ transition can be ascribed to changes in the surrounding of the Eu^{3+} ion, probably due to formation of europium phosphate. In their study of rare earth ions trimetaphosphates, Serra and Campos [39] observed that the $^5D_0 \rightarrow {}^7F_1$ transition is more intense than $^5D_0 \rightarrow {}^7F_2$. In recent works on double phosphates doped with rare earth ions [40], and ortho and metaphosphate doped with rare earth [41], the same behavior was detected for the emission spectra. Table 2 presents the relative intensities of the $^5D_0 \rightarrow {}^7F_0$ and $^5D_0 \rightarrow {}^7F_2$ transitions with respect to $^5D_0 \rightarrow {}^7F_1$ for all the samples.

The relative intensities of the bands observed for the samples are similar, since the surroundings of Eu^{3+} are the same. When these data are compared with those obtained for the samples containing no phosphate ions, it can be said that the Eu^{3+} symmetry increases in the presence of phosphate.

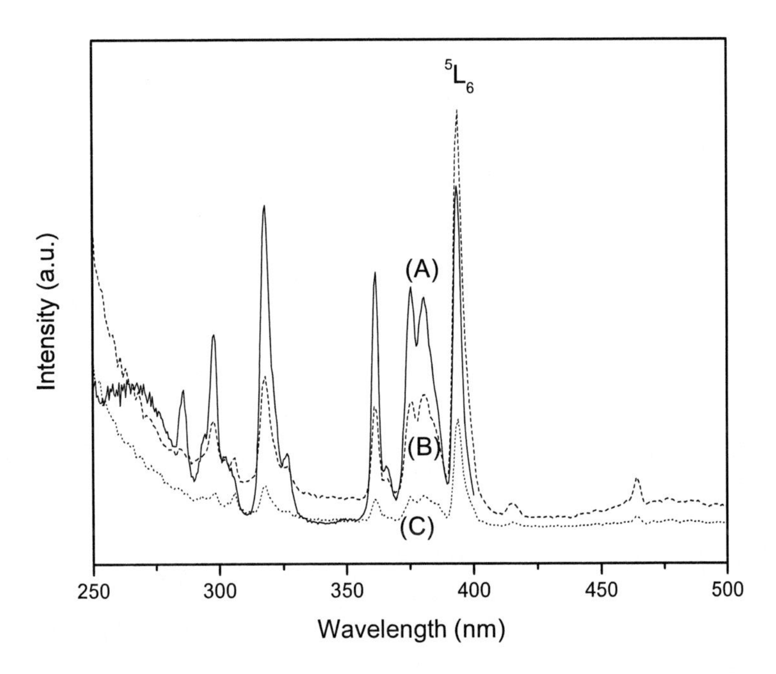

Figure 9. Excitation spectra of the Eu^{3+} ion doped into silica matrices containing different TEOS/PTES/$NH_4H_2PO_4$ molar ratios: (A) 1:0:0.25, (B) 0:1:0.25, and (C) 1:1:0.25.

Table 2. Relative intensity ratio of the $^5D_0 \rightarrow {}^7F_0$ / $^5D_0 \rightarrow {}^7F_1$ and $^5D_0 \rightarrow {}^7F_2$ / $^5D_0 \rightarrow {}^7F_1$ transitions

Xerogels	$^5D_0 \rightarrow {}^7F_0/{}^5D_0 \rightarrow {}^7F_1$	$^5D_0 \rightarrow {}^7F_2/{}^5D_0 \rightarrow {}^7F_1$
TEOS+Eu+$NH_4H_2PO_4$	0.04	1.17
PTES+Eu+$NH_4H_2PO_4$	0.03	0.98
TEOS+PTES+Eu+$NH_4H_2PO_4$	0.05	1.09

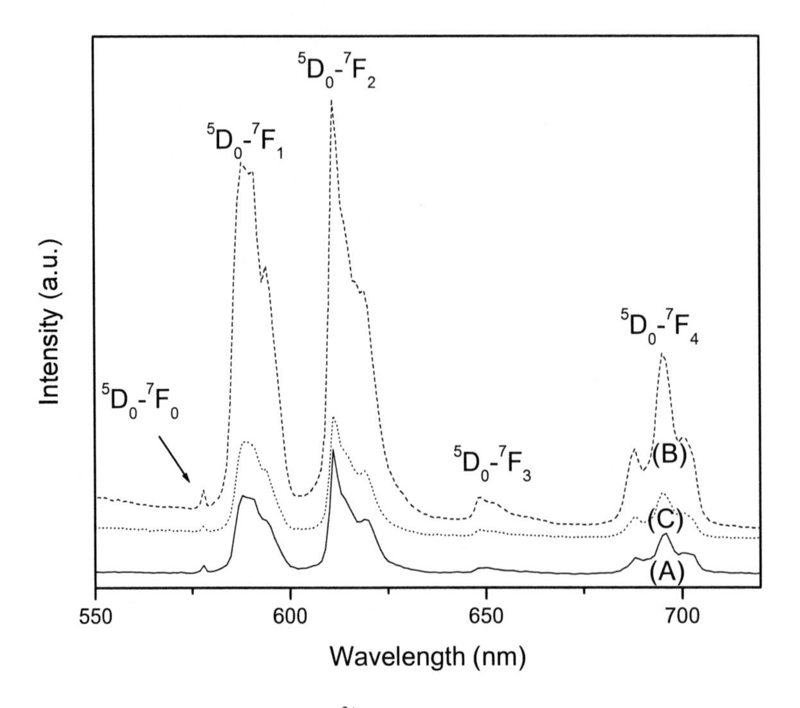

Figure 10. Emission spectra of the Eu^{3+} ion doped into silica matrices containing different TEOS/PTES/$NH_4H_2PO_4$ molar ratios: (A) 1:0:0.25, (B) 0:1:0.25, and (C) 1:1:0.25.

In order to prepare new hybrid materials by the hydrolytic sol-gel process, we used other alkoxides, so as to verify their influence on the photoluminescence of the Eu^{3+} ion. One new material was prepared using a 1:1 tetraethylorthosilicate (TEOS) 3-methacryloxypropyltrimethoxysilane (MPTMS) molar ratio. The solution was submitted to continuous magnetic stirring at room temperature, and a saturated ethanol ammonium solution was used as catalyst. After 40 minutes of stirring, the solvent was evaporated at room temperature for 1 week. The resulting material was heat-treated at 50, 100, 150, and 200°C for 4 hours.

Figure 11 displays the excitation spectra of the Eu^{3+} ion doped into the silica matrix prepared with the TEOS and MPTMS alkoxides. The maximum emission was found to occur at 612 nm ($^5D_0 \rightarrow {}^7F_2$). The band ascribed to the transition from $^7F_0$ (fundamental level) to $^5L_6$ (excitation level) was observed in three samples only, namely those treated at 100, 150, and 200°C. The samples treated at 25 and 50°C exhibited a wide band at 340 nm.

Compared with the silica matrix prepared with the TEOS and PTES alkoxides and submitted to different heat-treatment temperatures (Figure 5), a large band between 300 and 400 nm appears for the TEOS/MPTMS samples treated at 200 and 300°C. This may be due to the presence of the MPTMS alkoxide.

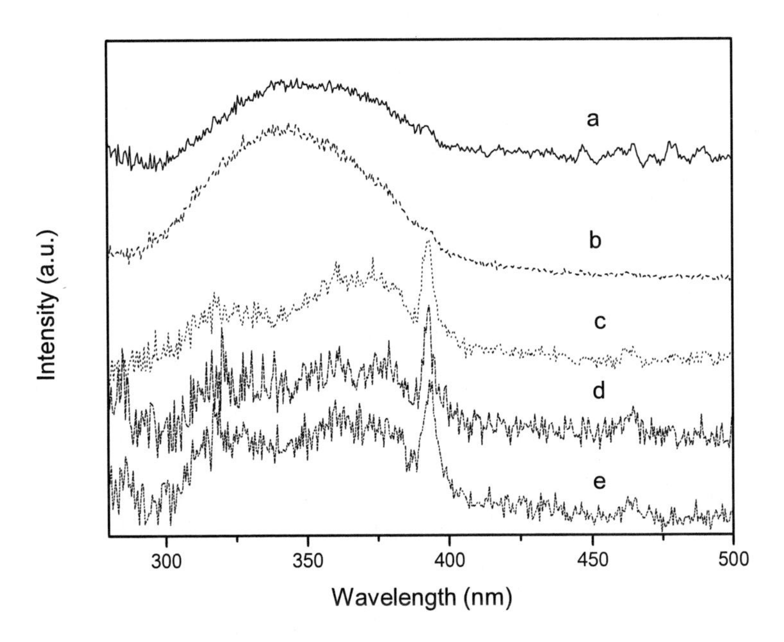

Figure 11. Excitation spectra of the Eu^{3+} ion doped into silica matrix prepared with the TEOS and MPTMS alkoxides, treated at the following temperatures: (a) 25°C, (b) 50°C, (c) 100°C, (d) 150°C, or e) 200°C.

Figure 12 shows the emission spectra of the Eu^{3+} ions doped into the matrix prepared with TEOS/ MPTMS, excited at $^5L_6$ (393 nm).

The emission spectra presented transitions arising from $^5D_0$ to ^7F_j (J = 0, 1, 2, 3 and 4) manifolds. The Eu^{3+} emission bands in these spectra are characterized by nonhomogenous distributions of the ion in the silica matrix [34 - 38]. The emission spectra are similar to those of the hybrid TEOS/PTES matrix treated at different temperatures, thereby indicating that Eu^{3+} is located in the silica matrix, and that the organosilane alkoxide has no influence on the ion's surrounding.

Table 3 presents the relative intensity of the $^5D_0 \rightarrow {}^7F_0$ and $^5D_0 \rightarrow {}^7F_2$ transitions with respect to the $^5D_0 \rightarrow {}^7F_1$ transition, for all the samples.

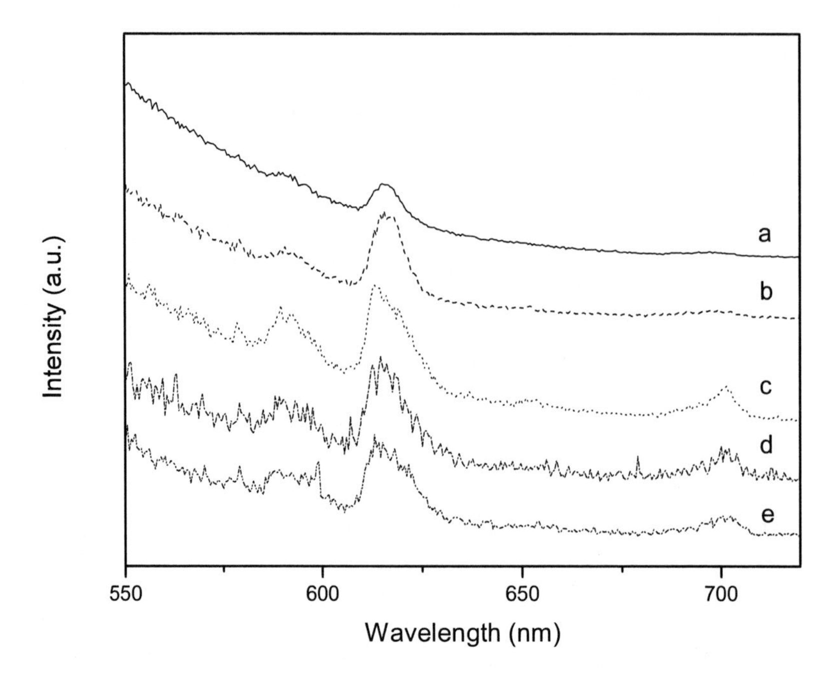

Figure 12. Emission spectra of the Eu^{3+} ion doped into silica matrix prepared with the TEOS and MPTMS alkoxides, treated at the following temperatures: (a) 25°C, (b) 50°C, (c) 100°C, (d) 150°C, or e) 200°C. $\lambda_{exc.}$ = 393 nm.

Table 3. Relative intensity ratio of the $^5D_0 \rightarrow {}^7F_0$ / $^5D_0 \rightarrow {}^7F_1$ and $^5D_0 \rightarrow {}^7F_2$ / $^5D_0 \rightarrow {}^7F_1$ transitions

Samples	$^5D_0 \rightarrow {}^7F_0/{}^5D_0 \rightarrow {}^7F_1$	$^5D_0 \rightarrow {}^7F_2/{}^5D_0 \rightarrow {}^7F_1$
TEOS+MPTMS (25°C)	0.23	1.41
TEOS+MPTMS (50°C)	0.23	1.70
TEOS+MPTMS (100°C)	0.23	1.34
TEOS+MPTMS (150°C)	0.23	1.51
TEOS+MPTMS (200°C)	0.23	1.82

The relative intensities of the bands corresponding to the $^5D_0 \rightarrow {}^7F_0$ / $^5D_0 \rightarrow {}^7F_1$ transition are the same for all heat-treatment temperatures. As for the $^5D_0 \rightarrow {}^7F_2$ / $^5D_0 \rightarrow {}^7F_1$ ratio, the heat-treatment temperature can affect the surrounding of the ion, due to decomposition of the organic molecules of the MPTMS alkoxide. Indeed, upon increasing temperature, the ^{29}Si NMR reveals the disappearance of the peaks at -60 and -50 ppm (chemical shift), ascribed to T^2 and T^1, respectively [12].

The weak emission intensity might be due to the presence of the –OH group in the silica matrix, which can lead to incomplete polymerization. The degree of hydrolysis and condensation of the silica matrix obtained by the sol-gel route can be controlled.by several parameters such as type of silane, temperature, catalyst, among other reaction conditions. In the present case, the use of the MPTMS alkoxide can affect the degree of hydrolysis and condensation, producing a large quantity of –OH groups in the matrix.

Our next subject will be the hybrid materials prepared with TEOS and MPTMS and the study of the influence of TEOS pre-hydrolysis on silica shape. Eu^{3+} was used as probe in this investigation.

Silica particles were prepared by the hydrolytic sol-gel route using the organofunctionalized alkoxides MPTMS and TEOS at a 1:1 molar ratio, or the alkoxide TEOS only. The silica was obtained from a mixture of water and isopropyl alcohol at a 1:1 molar ratio, using basic catalysis (ammonium hydroxide), under magnetic stirring at a temperature of around 20°C. To this end, 0.710 mL TEOS and 0.756 mL MPTMS were added to the solution. 1% Eu^{3+} ions, in ethanol solution 0.1 $mol.L^{-1}$, were added to the silica, to obtain structural information about the silica matrix. The $EuCl_3$ was prepared from europium (III) oxide (Eu_2O_3), calcined at 900°C for 2h. To this end, 0.8798g Eu_2O_3 was dissolved in HCl 6 $mol.L^{-1}$. The final ethanolic solution was 0.1 $mol.L^{-1}$ concentration. Six samples were produced with different TEOS pre-hydrolysis and condensation times. The first sample was prepared by adding MPTMS soon after TEOS addition, while four of the five remaining samples were prepared by waiting for 5, 15, 30, and 45 minutes before MPTMS addition, respectively; the fifth sample was prepared without MPTMS. The samples were centrifuged, washed in ethanol, and dried at 50°C for 1 day.

The characterizations revealed the effect of the TEOS alkoxide hydrolysis and condensation time and the moment of MTPMS addition. The TEM images of the samples to which MPTMS was added after 5, 15, 30, or 45 min. and of the sample containing no MPTMS indicated the formation of agglomerated, overlapping, and dense spherical nanoparticles of different sizes. A comparison of the sample without added MPTMS and the one prepared with the addition of MPTMS after 45 min. revealed similar morphology but different particle sizes. The sample prepared without the addition of MPTMS also presented silica nanoparticles, although the particles were smaller. An analysis by electron diffraction confirmed the amorphousness of all the samples.

The Eu^{3+} excitation and emission spectra recorded for all the samples are illustrated in Figures 13 and 14, respectively.

The band ascribed to the transition from $^7F_0$ (fundamental level) to $^5L_6$ (excitation level) was observed in the samples containing MPTMS added after 0, 5, 15, and 30 min. of TEOS addition, but it was absent from the samples containing MPTMS added after 45 min. and from the sample without MPTMS. Caiut et al [42] observed the O-Eu charge transfer state (CTS) in $SiO_2:Eu^{3+}$, which was not detected in our system.

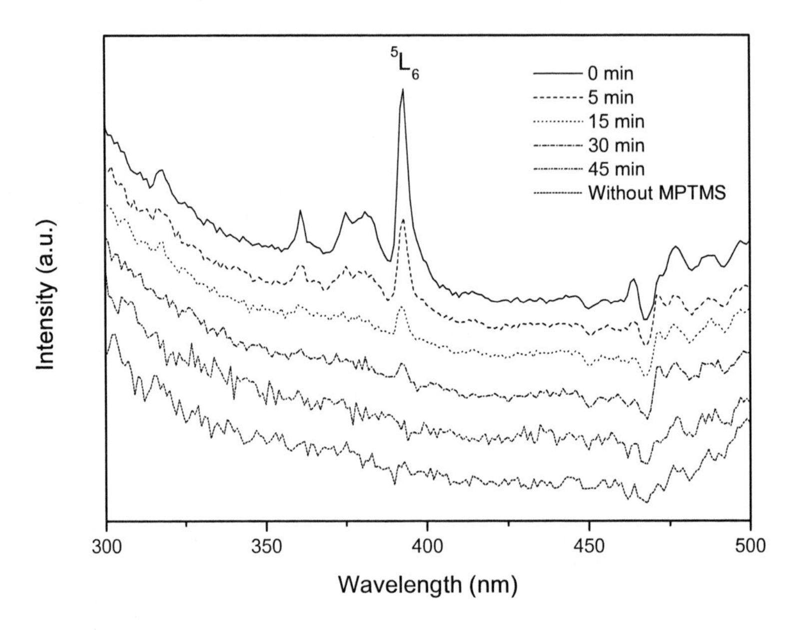

Figure 13. Excitation spectra of the Eu^{3+} ions in the silica matrix of the samples containing MPTMS added after 0, 5, 15, 30, and 45 min of TEOS addition, and of the sample without MPTMS, $\lambda_{em} = 612$ nm.

In the case of Figure 14, which shows the emission spectra of Eu^{3+} ions in the matrix; the excitation wavelength was recorded at 392 nm, $^5L_6$ level.

The luminescence spectra of Eu^{3+} ions in the hybrid materials evidence a broad emision in the blue region of the electromagnetic spectra. This band is usually observed for siloxanes and assigned to electron-hole recombination due to defects on the surface of siloxane particles [43]. The Eu^{3+} emission bands in these spectra were characterized by a nonhomogenous distribution of the ion in the silica matrix [34 – 36], and they are the same as those observed in the case of the PTES alkoxide (figure 6, 7 and 8). The typical emission bands of Eu^{3+} ions were observed in the regions of 580, 590, 612, 650, and 700 nm, corresponding to the electronic transitions from the excited level $^5D_0$ to the fundamental level $^7F_{J\ (J = 0, 1, 2, 3\ \text{and}\ 4)}$.

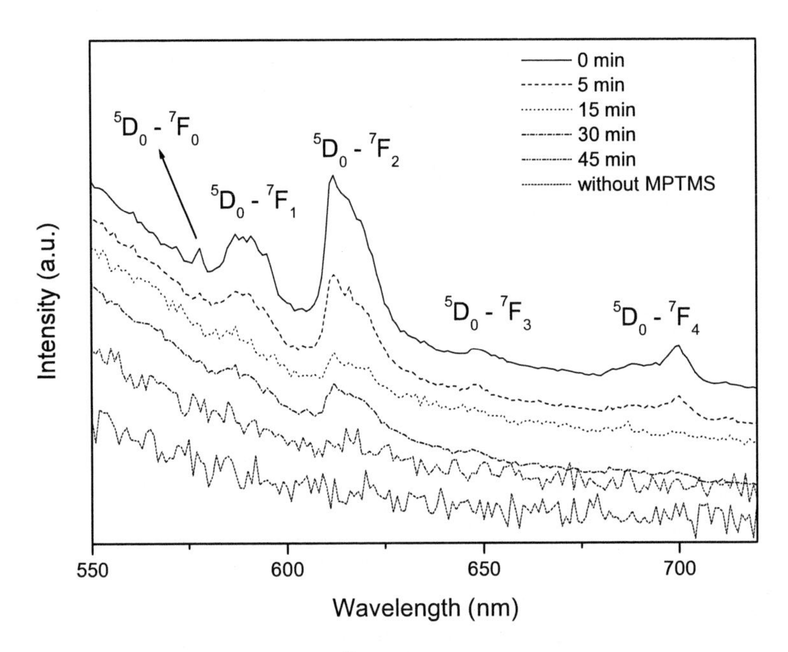

Figure 14. Emission spectra of the Eu^{3+} ions in the silica matrix of the samples containing MPTMS added after 0, 5, 15, 30, and 45 min. of TEOS addition, and of the sample without MPTMS, $\lambda_{exc} = 392$ nm.

The samples to which MPTMS was added after 0, 5, and 15 min displayed bands corresponding to the transition of the excitation level $^5D_0$ to the fundamental level $^7F_{J\ (J = 0, 1, 2, 3\ \text{and}\ 4)}$. The band in the region of 579 nm in the spectra of these samples suggest the presence of a site without inversion center, which can be confirmed by the presence of a $^5D_0 \rightarrow {}^7F_2$ transition band that should be absent from the spectrum of sites with inversion.

The intensity of the Eu^{3+} ion emission decreased as a function of the timing of MPTMS addition. This is probably due to the rise in the number of water molecules in the samples, as indicated by the thermogravimetric curve. A larger number of water molecules leads to loss of Eu^{3+} vibrational energy. However, it is more likely that most of the Eu^{3+} ions were eliminated during the washing process, since they were added together with MPTMS, after TEOS condensation. The sample containing no MPTMS did not display the typical band of the Eu^{3+} ion emission, probably because of energy loss by vibrational modes of the –OH groups on the silica surface, resulting from incomplete condensation of the TEOS alkoxide. Indeed, the presence of –OH groups directly interferes in the emission of the Eu^{3+} ion.

As mentioned earlier, because of the electric-dipole character of Eu^{3+}, the intensities of the $^5D_0 \rightarrow {}^7F_0$ and $^5D_0 \rightarrow {}^7F_2$ transitions are strongly dependent on the ion's surrounding [44]. The band corresponding to the $^5D_0 \rightarrow {}^7F_1$ transition, on the other hand, presents a magnetic dipole nature, so its intensity is not affected by the surroundings and it can thus be considered a standard to measure the relative intensity of the other bands [37]. The intensity ratios were lower for the hybrids, revealing that the Eu^{3+} ions are located in sites of lower symmetry, as observed in other works [45, 11]. Table 4 shows the relative intensity of the $^5D_0 \rightarrow {}^7F_0$ and $^5D_0 \rightarrow {}^7F_2$ transitions with respect to the $^5D_0 \rightarrow {}^7F_1$ transition for all the samples.

The relative intensity of the bands corresponding to the $^5D_0 \rightarrow {}^7F_2$ / $^5D_0 \rightarrow {}^7F_1$ transition decreases with the MPTMS addition time. This is an indication of increased symmetry around the Eu^{3+} ion and suggests that Eu^{3+} is not surrounded by any silica particles, which in turn are probably formed before MPTMS addition. In this case, hydrolysis of the precursor affected incorporation of the Eu^{3+} probe into the silica particles.

Table 4. Relative intensity ratio of the $^5D_0 \rightarrow {}^7F_0$ / $^5D_0 \rightarrow {}^7F_1$ and $^5D_0 \rightarrow {}^7F_2$ / $^5D_0 \rightarrow {}^7F_1$ transitions

Samples	$^5D_0 \rightarrow {}^7F_0/{}^5D_0 \rightarrow {}^7F_1$	$^5D_0 \rightarrow {}^7F_2/{}^5D_0 \rightarrow {}^7F_1$
TEOS+MPTMS (added at 0 min)	0.29	1.38
TEOS+MPTMS (added at 5 min)	0.37	1.26
TEOS+MPTMS (added at 15 min)	0.33	1.12
TEOS+MPTMS (added at 30 min)	0.34	1.08
TEOS+MPTMS (added at 45 min)	-	-
TEOS (without MPTMS)	-	-

The band at 1632 cm^{-1} in the infrared spectra of the samples can be ascribed to water bending. Vibrations in the 3000-3750 cm^{-1} region were attributed to -OH stretching on the silica surface and also to water left over from the synthesis, indicating incomplete TEOS condensation. The band related to Si-OH group vibrations was observed in the region of 946 cm^{-1} [46, 47]. It is noteworthy that the intensity of this band increases as a function of MPTMS addition time, indicating greater TEOS hydrolysis and incomplete condensation. According to Vinod et al., the band at 2350 cm^{-1} in organosilane polymers represents defects on the silica surface [48]. The intensity of this band was low. The bands corresponding to the organic matter, namely C=O at 1711 cm^{-1}, C-COO^- at 1306 cm^{-1}, CH_2= at 1397 cm^{-1}, and CH- at 2940 cm^{-1} [49, 50], had their intensities reduced as a function of MPTMS addition time, thereby confirming the thermal analysis results that indicated the presence of –OH groups in the silica matrix. All the samples contained residual organic matter originating from either MPTMS or incomplete TEOS hydrolysis. However, the amount of MPTMS-related organic matter decreased in the 5 and 15 min samples and was absent from the 30 and 45 min samples. This indicates that a longer wait prior to MPTMS addition may have allowed for greater separation of the organic component and for a larger amount of MPTMS-related organic matter to be eliminated during material washing. The scheme 2 can be represents the hybrid structure.

In this work, another silica sample was prepared in ethanol solvent at a 3:1:7 tetraethylorthosilicate (TEOS)/3aminopropyltriethoxysilane (APTS)/distilled water (H_2O) molar ratio. Acetic acid was used as catalyst. The Eu^{3+} chloride or Eu^{3+} 2,2′-bipyridine compounds were added to the sol (TEOS/APTS/H_2O) under stirring at room temperature. Ethanol was evaporated at ~25°C, producing a white solid. The powder was dried at 50°C overnight.

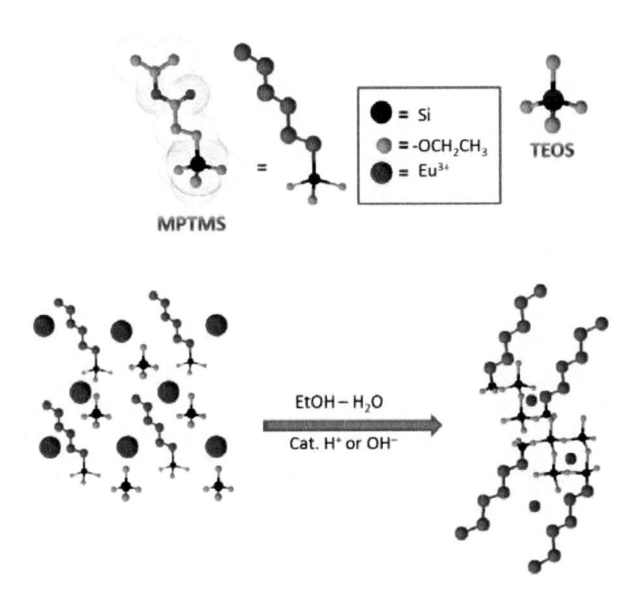

Scheme 2. Schematic representation of the hybrid sctructure contends MPTMS alkoxide.

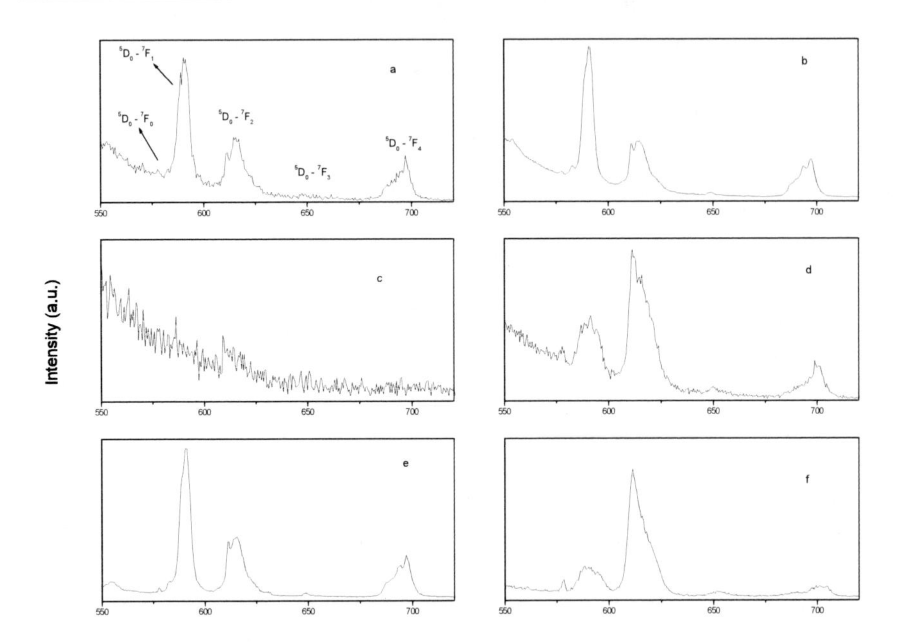

Figure 15. Emission spectra for the samples (a) $TEOSEuCl_3$ λ_{exc} = 393 nm, (b) $TEOSEu(bpy)Cl_3$ λ_{exc} = 393 nm, (c) $TEOSAPTSEuCl_3$ λ_{exc} = 393 nm, (d) $TEOSAPTSEu(bpy)Cl_3$ λ_{exc} = 393 nm, (e) $TEOSEu(bpy)Cl_3$ λ_{exc} = 320 nm, (f) $TEOSAPTSEu(bpy)Cl_3$ λ_{exc} = 311 nm.

Control samples were prepared from TEOS and TEOS+APTS, hence silica particles (TEOS) and modified silica particles (TEOS/APTS), which were used as standards. The results for all the samples were compared to those achieved for these control particles, so as to better understand their behavior. The infrared spectra of the TEOS/APTS samples are different with respect to the bands at wavenumbers 956, 1600, and 2939 cm^{-1}, ascribed to Si-OH, NH_2, and CH, respectively [51]. The vibrational mode at 1600 cm^{-1} appears in all the samples containing aminopropyl groups.

The excitation spectra of the samples containing the compound $Eu(bpy)Cl_3$ present a new broad band with maximum at about 310 nm. This maximum can be attributed to the absorption maxima of the ligand bpy. Figure 15 corresponds to the emission spectra of the Eu^{3+} ion in the various samples, excited at 393 nm ($^5L_6$ level of Eu^{3+} and ligand band).

For the first time in this work, the emission spectra of the Eu^{3+} ions are different from those observed so far. The band corresponding to the $^5D_0 \rightarrow {}^7F_1$ transition is relatively more intense than that due to the $^5D_0 \rightarrow {}^7F_2$ transition.

In the samples that do not contain the modifying agent (APTS), the relative intensity between the $^5D_0 \rightarrow {}^7F_1$ magnetic dipole transition (~591 nm) and the $^5D_0 \rightarrow {}^7F_2$ electric dipole transition (~612 nm) in the emission spectra suggests a local symmetry for the Eu^{3+} ion with inversion center [52]. Water-like environment for the Eu^{3+} ions has been observed for these systems [53], showing that Eu^{3+} ions can be located in porous silica. This behavior was not observed for the samples containing the modifying agent APTS, so the Eu^{3+} compounds are probably coordinated to the nitrogen atom of the amino propyl groups, indicating a longer distance between the Eu^{3+} ion and the silica surface, as represented in schemes 3 and 4. This behavior had been observed by us in functionalized commercial silica gel containing imidazole propyl groups [54].

Table 5 shows the relative intensity of the $^5D_0 \rightarrow {}^7F_0$ and $^5D_0 \rightarrow {}^7F_2$ transitions with respect to the $^5D_0 \rightarrow {}^7F_1$ transition for all the samples.

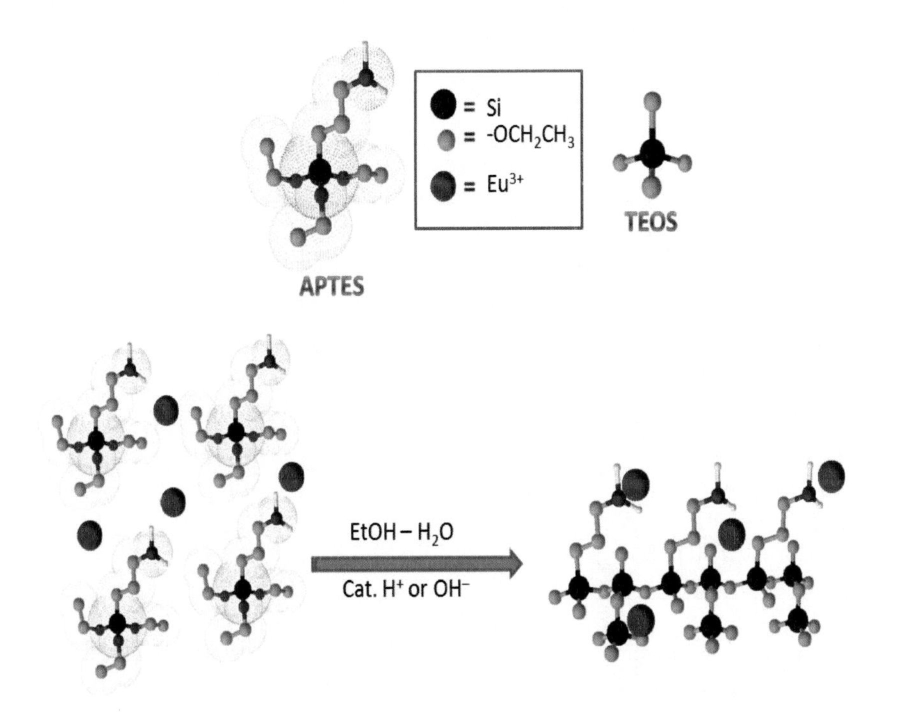

Scheme 3. Schematic representation of the hybrid sctructure contends APTS alkoxide doped with $EuCl_3$

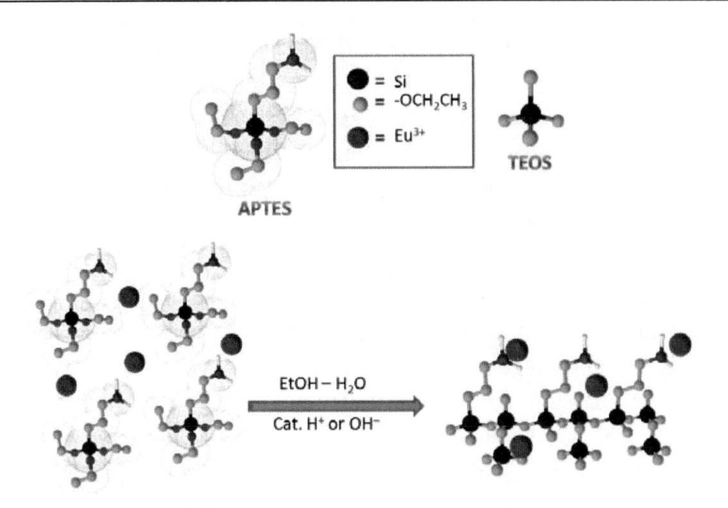

Scheme 4. Schematic representation of the hybrid sctructure contends APTS alkoxide doped with Eu^{3+} compounds.

Table 5. Relative intensity ratio of $^5D_0 \rightarrow {}^7F_0 / {}^5D_0 \rightarrow {}^7F_1$ and $^5D_0 \rightarrow {}^7F_2 / {}^5D_0 \rightarrow {}^7F_1$ transitions

Samples	$^5D_0 \rightarrow {}^7F_0/{}^5D_0 \rightarrow {}^7F_1$	$^5D_0 \rightarrow {}^7F_2/{}^5D_0 \rightarrow {}^7F_1$
TEOS+$EuCl_3$ (λ_{exc} = 393 nm)	0.09	0.65
TEOS+Eu(bpy)Cl_3 (λ_{exc} = 393 nm)	0.08	0.61
TEOS+APTS+$EuCl_3$ (λ_{exc} = 393 nm)	0.27	0.94
TEOS+APTS+Eu(bpy)Cl_3 (λ_{exc} = 393 nm)	0.19	1.80
TEOSEu(bpy)Cl_3 (λ_{exc} = 320 nm)	0.03	0.67
TEOSAPTSEu(bpy)Cl_3 (λ_{exc} = 311 nm)	0.13	3.33

Bearing in mind that the $^5D_0 \rightarrow {}^7F_2$ emission is due to an electric dipole transition and is particularly dependent upon the local symmetry [44, 55], whereas the $^5D_0 \rightarrow {}^7F_1$ emission is allowed by magnetic dipole considerations, so it is indifferent to the local symmetry, the ratio between $^5D_0 \rightarrow {}^7F_2 / {}^5D_0 \rightarrow {}^7F_1$ emission intensities therefore gives us valuable information about changes to the environment around the Eu^{3+} ion. The high value obtained for the $^5D_0 \rightarrow$

$^7F_2$ / $^5D_0 \rightarrow {}^7F_1$ relative intensity in the case of the samples containing the agent APTS indicates that Eu^{3+} is situated at low symmetry sites.

The Eu^{3+} emission could be excited by the antenna effect through the bpy molecules, where the ligand absorbs and transfers energy to the Eu^{3+} ion.

Knowing that the sol-gel process can be used to prepare many kinds of materials with several applications, in the next part we will describe the use of this methodology to prepare materials with potential applications in biomaterials. The Eu^{3+} ion was used as structural probe.

In this stage of the study, materials containing Ca-P-Si were prepared by the sol-gel methodology; tetraethylorthosilicate (TEOS), calcium alkoxide, and phosphoric acid were used as precursors. The material was synthesized in 10 mL ethanol (solvent), to which were added $9.60x10^{-3}$ mol TEOS, $3.84x10^{-3}$ mol calcium ethoxide, and $6.73x10^{-4}$ mol phosphoric acid, under stirring and basic catalysis (NH_3 ethanolic saturated solution). $EuCl_3$ in ethanolic solution was added as structural probe. After 5 hours a gel was formed, and the bulk was obtained after drying at 50°C for 1 day. The resulting material was immersed in Simulated Body Fluid (SBF) [56], pH = 7.40, for 12 days. The sample was characterized before and after contact with SBF.

Figure 16 depicts the excitation spectra of the Ca-P-Si matrix doped with Eu^{3+} ions by the sol-gel method. The maximum emission was found to occur at 612 nm ($^5D_0 \rightarrow {}^7F_2$). Note the band from $^7F_0$ (fundamental level) to $^5L_6$ (excitation level).

Figure 17 corresponds to the emission spectra of the Ca-P-Si matrix doped with Eu^{3+} ions, showing the excitation wavelength at the $^5L_6$ level (394 nm).

The emission spectra presented transitions arising from $^5D_0$ to ^7F_J (J = 0, 1, 2, 3 and 4) manifolds. The Eu^{3+} emission bands in these spectra are characterized by nonhomogenous distributions of the ion in the Ca-P-Si matrix.39-41. The excitation and emission spectra are similar to those observed for the hybrid samples prepared with TEOS/PTES/phosphate ion (Figure 10). This is an indication that europium phosphate can be formed. Transmission electron microscopy revealed the formation of small particles with an average size of 20 nm, and energy dispersive X-ray spectroscopy indicated that the materials have an amorphous phase with large quantities of Si and O, amorphous silicate, and lower quantities of Eu and Cl. The emission spectra of the Eu^{3+} ions demonstrate the presence of a non-homogeneous surrounding site, charaterizing an amorphous phase. The crystalline phase, whose composition contained Ca, P, and Eu, was ascribed to crystallization of hydroxyapatite. The electron diffraction of this phase revealed distances of 2.86 and 1.88 Å, which indicates that these distances correspond to $2\theta = 31.2°$

(211) and 48.6° (320) according to Bragg´s law. These peaks were ascribed to hydroxyapatite (JCPDS – 9-0432) [57].

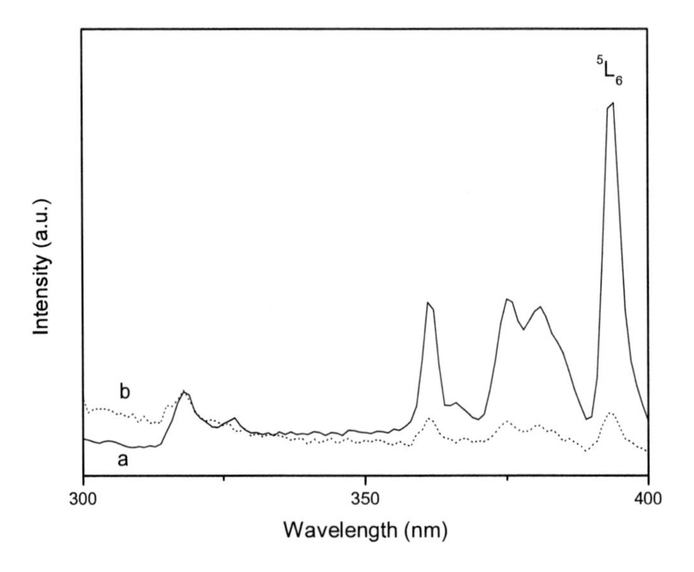

Figure 16. Excitation spectra of the Ca-P-Si matrix doped with Eu^{3+} ions before (a) and after (b) contact with SBF, $\lambda_{em.} = 612$ nm.

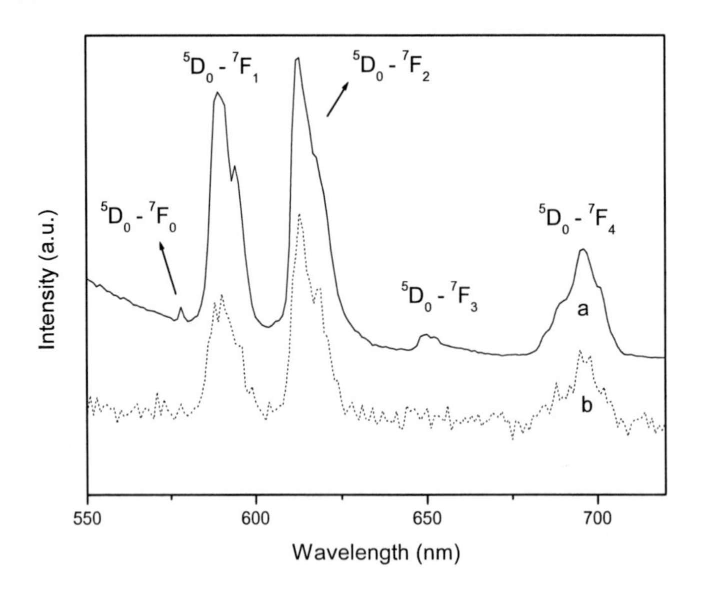

Figure 17. Emission spectra of the Ca-P-Si matrix doped with Eu^{3+} ions before (a) and after (b) contact with SBF, $\lambda_{exc.} = 394$ nm.

Table 6 shows the relative intensity of the $^5D_0 \rightarrow {^7F_0}$ and $^5D_0 \rightarrow {^7F_2}$ transitions with respect to the $^5D_0 \rightarrow {^7F_1}$ transition. The gradual increase in the relative ratio of the transitions indicates a small change in the surrounding of the Eu^{3+} ions after contact with the SBF solution. The relative intensity is similar to those achieved for the hybrid samples containing TEOS/PTES/phosphate ions (figure 10), suggesting that addition of phosphate ions promotes formation of europium phosphate.

To continue our studies on materials prepared by sol-gel containing phosphate ions, for application in biomaterials, we prepared samples with Ca/P molar ratios equal to or different from that encountered in natural hydroxyapatite (HA). The sample designated HA= had the same Ca/P molar ratio as HA, the HA- sample had lower Ca/P molar ration and the HA+ sample had larger Ca/P molar ratio than HA. The three samples, HA=, HA+, and HA-, were synthesized by the sol-gel route. To this end, 9.60×10^{-3} mols TEOS, 3.84×10^{-3} mols calcium ethoxide, and 6.06×10^{-4} (HA-), 6.73×10^{-4} (HA=) or 7.40×10^{-4} (HA+) mols phosphoric acid were added to 16.0 mL ethanol (solvent), under stirring. The resulting materials were dried at 50°C for 1 day. 1% $EuCl_3$ in ethanolic solution was added before gelation, since the Eu^{3+} ion can replace the Ca^{2+} ion and hence be used as structural probe. The materials were then immersed in SBF pH = 7.40 at 37°C. To this end, 0.10 g of the powder sample was placed into 10 mL SBF solution for 19 days. The samples were characterized before and after contact with SBF.

Figures 18 and 19 respectively display the excitation and emission spectra of the samples HA=, HA+, and HA- prepared by the sol-gel method, doped with Eu^{3+} ions, before contact with the SBF solution. The maximum emission occurs at 590 nm ($^5D_0 \rightarrow {^7F_1}$), while the maximum excitation takes place at 394 nm ($^5L_6$ level).

Table 6. Relative intensity ratio of the $^5D_0 \rightarrow {^7F_0} / {^5D_0} \rightarrow {^7F_1}$ and $^5D_0 \rightarrow {^7F_2} / {^5D_0} \rightarrow {^7F_1}$ transitions and lifetime of the $^5D_0 \rightarrow {^7F_2}$ transition

Samples	5D0 → 7F0 / 5D0 → 7F1	5D0 → 7F2 / 5D0 → 7F1
Before SBF	0.07	1.15
After 12 days in SBF	0.13	1.29

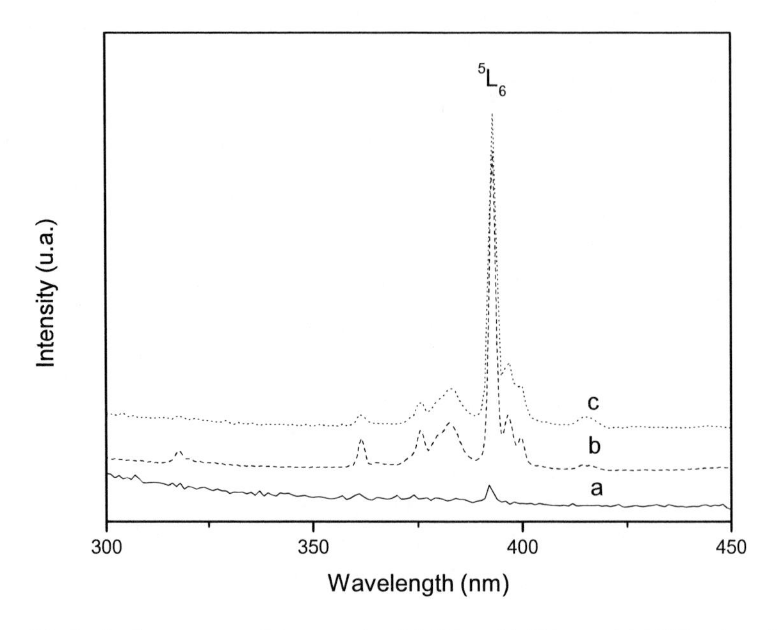

Figure 18. Excitation spectra of the samples (a) HA=, (b) HA-, and (c) HA+ doped with Eu^{3+} ions, before contact with the SBF solution, $\lambda_{em.}$ = 590 nm.

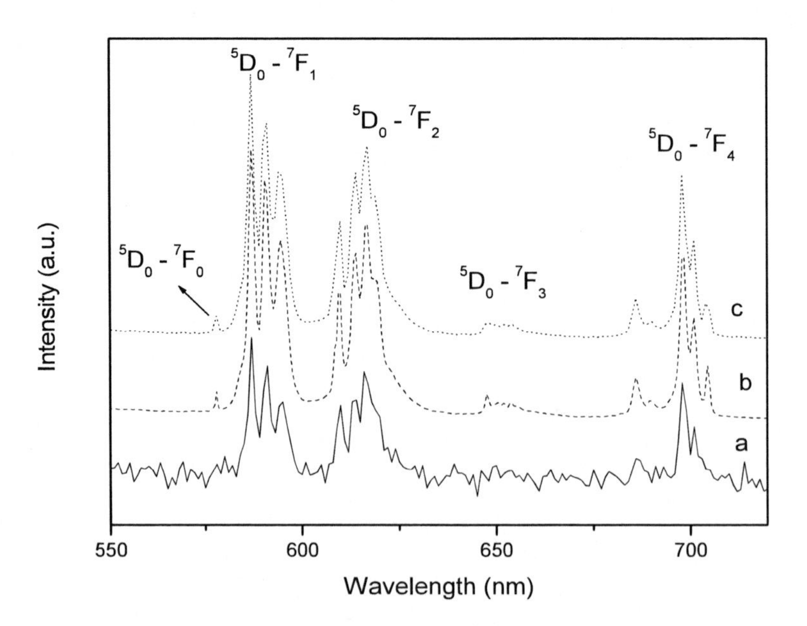

Figure 19. Emission spectra of the samples (a) HA=, (b) HA-, and (c) HA+ doped with Eu^{3+} ions before contact with the SBF solution, $\lambda_{exc.}$ = 394 nm.

The emission spectra reveal transitions arising from $^5D_0$ to ^7F_J (J = 0, 1, 2, 3, and 4) manifolds. There are three main transitions, namely $^5D_0 \rightarrow {}^7F_0$ (around 570 nm), $^5D_0 \rightarrow {}^7F_1$ (around 595 nm), and $^5D_0 \rightarrow {}^7F_2$ (around 610 nm). The first is a strongly forbidden transition not yet observed with appreciable intensity in some hosts. The $^5D_0 \rightarrow {}^7F_1$ transition is forbidden as electric dipole, but allowed as magnetic dipole. This is the only transition that takes place when Eu^{3+} occupies a site coinciding with a symmetry centre. When the Eu^{3+} ion is situated at a site lacking an inversion centre, the transition corresponding to even J values (except 0) are electric-dipole allowed, and red emission can be observed. The $^5D_0 \rightarrow {}^7F_1$ transition can also be detected as a magnetic-dipole allowed transition. Further on, all the lines corresponding to this transition split into a number of components depending on the local symmetry [58].When Eu^{3+} is located at a low symmetry site (without an inversion center), the $^5D_0 \rightarrow {}^7F_2$ emission transition often predominates in the emission spectrum, so the Eu^{3+} ions occupy a non-inversion symmetric site [59].The $^5D_0 \rightarrow {}^7F_0$ transition of Eu^{3+} is only allowed in the case of C_s, C_n, and C_{nv} symmetries of the Eu^{3+} sites in the crystalline state [60]. The hexagonal unit-cell HA contains ten cations distributed between two crystallographic sites: four on type (1) sites and six on type (2) sites. Ca (1) ions present C_3 symmetry and are surrounded by nine oxygen atoms. Ca (2) ions present Cs symmetry and are surrounded by six oxygen atoms [61 - 63]. There are literature reports stating that Ca (1) and Ca (2) ions (r = 0.99 Ǻ) can be replaced with Pb^{2+} II (r = 1.2 Ǻ) [63], so the Eu^{3+} ion (r = 0.95Ǻ) [65] can also occupy the Ca^{2+} sites in the crystalline hydroxyapatite and β-tricalcium phosphate (β-TCP), as well as in amorphous silica. The emission spectra of Eu^{3+} are not similar when the ion is located in glassy hosts, as discussed in the literature [66 - 68]. Table 7 shows the intensity of the $^5D_0 \rightarrow {}^7F_0$ and $^5D_0 \rightarrow {}^7F_2$ transitions relative to the $^5D_0 \rightarrow {}^7F_1$ transition.

The similar relative intensities of the transitions indicate that the environment around the Eu^{3+} ion is similar in the three samples, and the ion can be located either in the silica matrix or in calcium phosphate. The relative intensities are similar to those of other hybrid materials also containing phosphate ions and prepared by the sol-gel route (Figures 10 and 17).

The X-ray diffraction patterns of the samples prepared with different Ca/P molar ratios, before contact with the SBF solution, give evidence of crystalline and amorphous phases, with well-defined diffraction peaks. The crystalline phase displays peaks at 2θ = 26.5, 32.5, 33.0, 49.2, and 53.1°, which can be ascribed to hydroxyapatite (HA), whereas the peaks corresponding to calcium triphosphate (β–TCP-) appear at 2θ = 26.5, 30.2, and 53.1° [69]. Several other peaks due to other phosphate silicates, such as $Ca_5(PO_4)_2SiO_4$ and $(Ca_2(SiO_4))_6(Ca_3(PO_4)_2)$, can also be observed.

The characterized samples were then placed into SBF solution for 19 days, followed by characterization using the same techniques mentioned above.

The XRD patterns of the samples immersed in SBF display the same peaks observed for the samples not submitted to contact with SBF solution. Therefore, the crystalline phase formed during the sol-gel process remains intact in the silica matrix after sample immersion into SBF. TEM analysis also reveals that the crystalline phase does not disappear after contact with the SBF solution, and the presence of calcium phosphate nanoparticles in the samples is confirmed by this technique.

The photoluminescence data demonstrate that the excitation and emission spectra of the samples HA=, HA+, and HA- present the same behaviour as that detected for the samples before contact with the SBF solution.

Table 8 depicts data on the intensity of the $^5D_0 \rightarrow {}^7F_0$ and $^5D_0 \rightarrow {}^7F_2$ transitions relative to the $^5D_0 \rightarrow {}^7F_1$ transition.

Table 7. Relative intensity of the $^5D_0 \rightarrow {}^7F_0$ and $^5D_0 \rightarrow {}^7F_2$ transitions with respect to the $^5D_0 \rightarrow {}^7F_1$ transition in the emission spectra of the samples HA=, HA+, and HA- doped with Eu^{3+}, before contact with SBF

Samples	$^5D_0 \rightarrow {}^7F_0 / {}^5D_0 \rightarrow {}^7F_1$	$^5D_0 \rightarrow {}^7F_2 / {}^5D_0 \rightarrow {}^7F_1$
HA =	0.09	1.13
HA-	0.09	1.06
HA+	0.11	1.11

Table 8. Relative intensity of the $^5D_0 \rightarrow {}^7F_0$ and $^5D_0 \rightarrow {}^7F_2$ transitions with respect to the $^5D_0 \rightarrow {}^7F_1$ transition in the emission spectra of the samples HA=, HA+, and HA- doped with Eu^{3+}, after contact with SBF

Samples	$^5D_0 \rightarrow {}^7F_0 / {}^5D_0 \rightarrow {}^7F_1$	$^5D_0 \rightarrow {}^7F_2 / {}^5D_0 \rightarrow {}^7F_1$
HA=	0.05	0.99
HA-	0.09	1.11
HA+	0.12	1.15

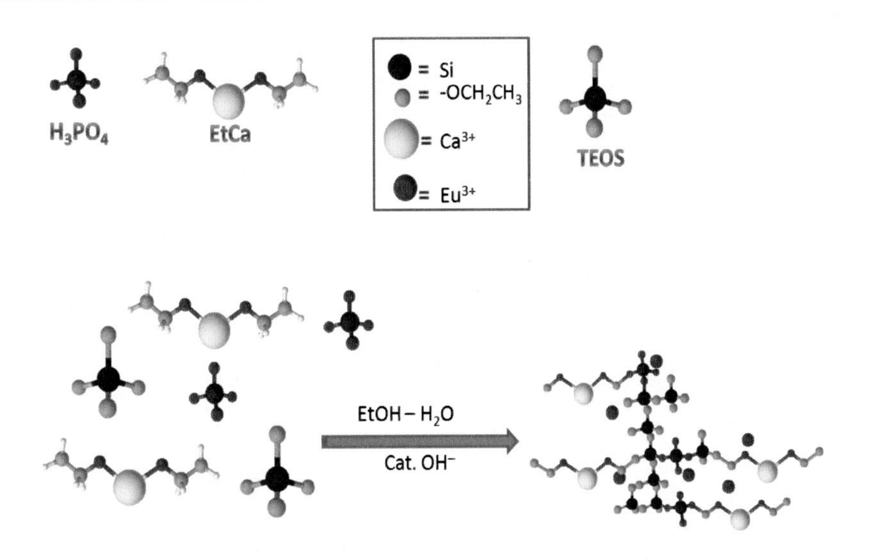

Scheme 5. Schematic representation of the silica particles obtained with Ca-Si-P.

There were only slight changes in the relative intensity of the emission bands, indicating that the environment around the Eu^{3+} ion is not affected upon contact of the samples with the SBF solution. This is because Eu^{3+} probably occupies a stable site in the matrix and in the phosphate nanoparticles. The scheme 5 presents a structure of the materials

In another stage of the present study we used the sol-gel process to prepare titanium oxide, monitored by using Eu^{3+} ion as luminescent probe. The titanium sol was prepared from tetraethylorthotitanate (TEOT) in ethanol (EtOH), and the metal alkoxide reaction was controlled by beta-diketone 2,4 pentanedione (acac) at a 1:1 molar ratio. To this end, 1.0 mmol acac was added to 10 mL EtOH under magnetic stirring. After 5 minutes, 1.0 mmol TEOT was added to the mixture. The ethanolic $EuCl_3$ solution was added to the sol at molar percentages of 0.1, 0.2, and 0.3%, and the sol was homogenized by magnetic stirring for 30 minutes. These solutions were dried at room temperature, and the resulting xerogels (powders) were heat-treated at 500, 750, and 1000°C, in porcelain crucibles.

Figures 20 and 21 illustrate the excitation and emission spectra of a sample doped with a molar percentage of 0.3% Eu^{3+} ion, treated at different temperatures.

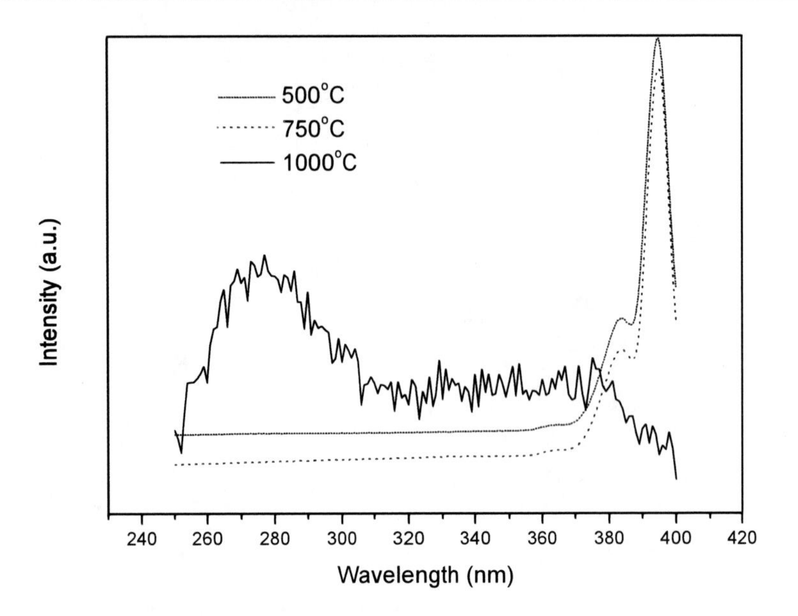

Figure 20. Excitation spectra of Eu^{3+} ions (0.3%) entrapped into the TiO_2 xerogel heat-treated at different temperatures.

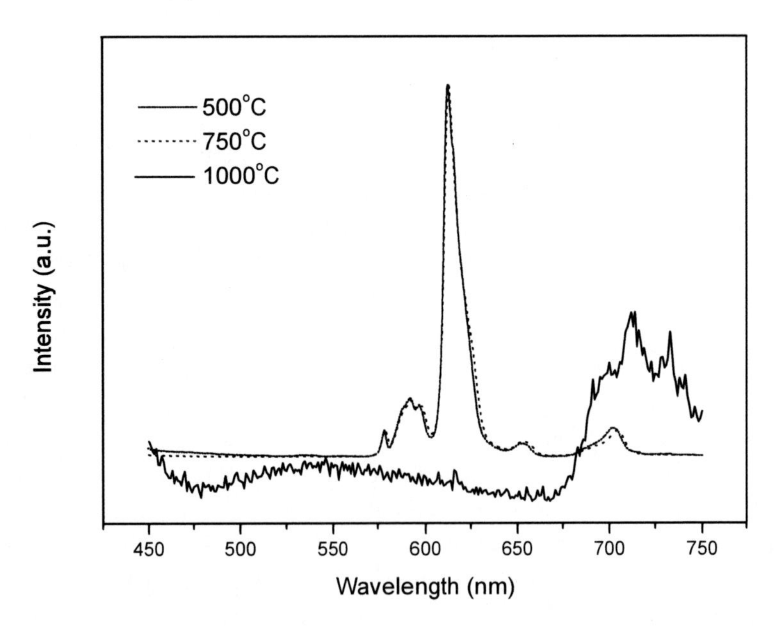

Figure 21. Emission spectra of Eu^{3+} ions (0.3%) entrapped into the TiO_2 xerogel heat-reated at different temperatures.

In the excitation spectra of the samples heated at 500 and 700°C, the maximum at 394 nm was ascribed to the $^5L_6$ level of Eu^{3+}. However, this band was not observed in the sample heated at 1000°C. The emission spectra presented transitions arising from $^5D_0$ to ^7F_j (J = 0, 1, 2, 3 and 4) manifolds excited at their maximum in the case of the samples heated at 500 and 700°C, but the sample heated at 1000°C showed no Eu^{3+} luminescence.

In the emission spectra of the samples treated at 500 and 700°C, the bands corresponding to the $^5D_0 \rightarrow {}^7F_2$ transition were more intense than those corresponding to the $^5D_0 \rightarrow {}^7F_1$ transition. This difference indicates that the Eu^{3+} occupies sites without inversion center [70 - 72]. The presence of $^5D_0 \rightarrow {}^7F_0$ transitions indicates that Eu^{3+} is located in a site with a C_{nv}, C_n, or C_s symmetry [73]. The presence of nonhomogeneous sites in the TiO_2 structure was observed based on the band-width emission [34]

There were only slight changes in the relative intensity of the emission bands for the Eu^{3+} ion located in the TiO_2 matrices, indicating that the environment around the Eu^{3+} ion is the same in these systems. When we compared these samples with the samples prepared with silicon alkoxide, we observed a larger ratio the between transitions $^5D_0 \rightarrow {}^7F_0$ / $^5D_0 \rightarrow {}^7F_1$. This indicates that the Eu^{3+} ion occupies a low symmetry site in the TiO_2 support, provided by the titania matrix

Table 9. Relative intensity of the $^5D_0 \rightarrow {}^7F_0$ and $^5D_0 \rightarrow {}^7F_2$ transitions with respect to the $^5D_0 \rightarrow {}^7F_1$ transition in the emission spectra into TiO_2 matrices

Samples	5D0 → 7F0 / 5D0 → 7F1	5D0 → 7F2 / 5D0 → 7F1
500°C	0.35	1.98
750°C	0.34	2.20
1000°C	-	-

The luminescence spectra of Eu^{3+} ions doped into TiO_2 xerogels and heat-treated at different temperatures (500 and 750°C) display similar luminescence bands. However, from 1000°C and above, the emission of Eu^{3+} ions disappears, probably due to a phase transition from anatase to rutile and to ion migration to the external surface. Clustering of the Eu^{3+} ion and quenching by energy transfer is observed. Indeed, the cluster was confirmed by X-ray diffraction and transmission electron microscopy. A large quantity of Eu^{3+} appeared on the surface of the titanium oxide heated at 1000°C, as observed by

Energy Dispersive X-ray Analysis (EDX), and quenching occured due to the high Eu^{3+} concentration. The transfer of emissive center energy to the level of another center promoted energy loss. The scheme 6 shows the structure of the titania matrix obtained by sol-gel, after heat treatment.

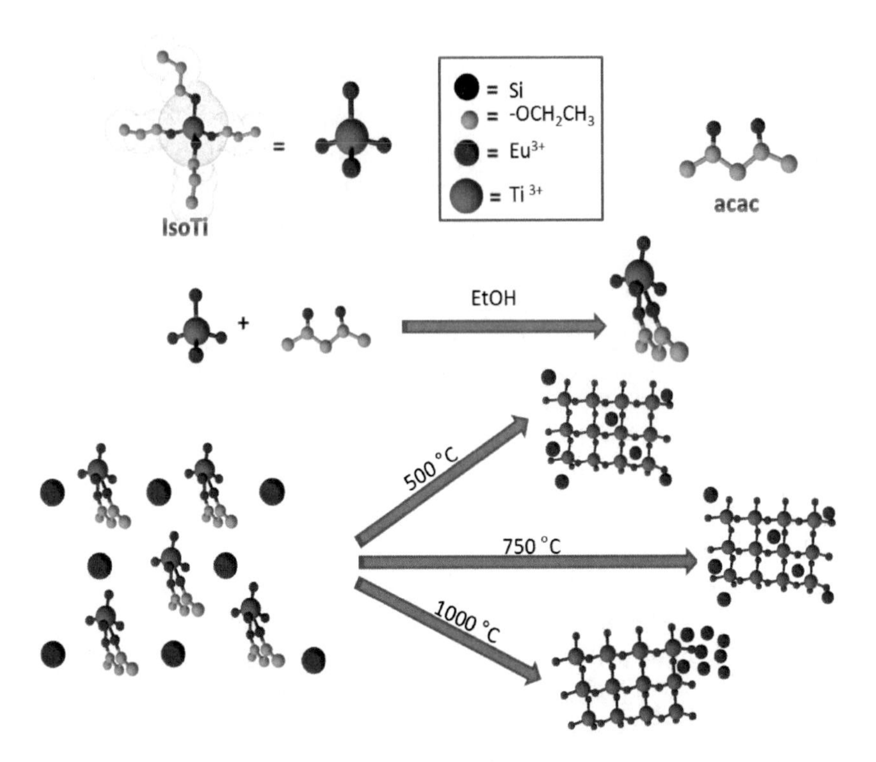

Scheme 6. Schematic representation of the titania particles obtained by sol-gel.

X-ray powder diffraction measurements indicated that the differences between percentages of Eu^{3+} were insignificant. The X-ray diffraction patterns showed that the material was totally amorphous at room temperature. Crystallization started at 500°C and, at 750°C, the sample had an almost totally crystalline appearance. After further elevation of the treatment temperature to 1000°C, the material presented a totally crystalline structure [74].

The TGA/DTA/DSC analyses gave evidence of a mass loss between 50 and 250°C, and another loss between 350 and 450°C. A structural change occurred from 450 to 750°C and from 950 to 1100°C, with DTA revealing two exothermic peaks at 450 – 750°C and an endothermic peak at 1000°C.

TEM for the xerogels treated at 500°C showed that the agglomerate displayed diameter not larger than 50 nm, with crystallization beginning in the anatase phase, and the EDX measurements were congruent with TiO_2 containing a low percentage of Eu^{3+}, probably due to the symmetry. The sample treated at 750°C had a diameter of about 35 nm and was almost entirely crystalline in the anatase phase. The xerogel presented a mixture similar to that of the sample treated at 500°C. TEM image of the sample treated at 1000°C presented a completely crystalline structure in rutile phase, as confirmed by X-ray powder diffraction and the selected area diffraction pattern (SADP). The EDX measurements revealed the consistency of titanium oxide with a high percentage of Eu^{3+}. In the rutile phase, Eu^{3+} ions migrated to the external surface of the material. This formation was responsible for quenching the luminescence of the Eu^{3+} ions and was probably caused by the transition from anatase to rutile phase, as confirmed by DSC and X-ray powder diffraction.

The versatility of the sol-gel process to produce different materials with various forms led us to use sol containing titanium alkoxide to prepared films. Sols were prepared by mixing TEOT (1.0×10^{-3} mol/L) in dry ethanol, followed by addition of the beta-diketonate 2,4-pentanedione at a 1:1 molar rtion. The ethanolic $EuCl_3$ was added to the sol, at different molar ratios. A yellow transparent sol was obtained. The sol was vigorously homogenized by magnetic or ultrasonic stirring for 15 minutes. The films were obtained by the dip-coating technique. Borosilicate glass substrates were carefully cleaned and then sunk into the sols. They were withdrawn at a rate of 100, 200, or 300 mm/min, and the resulting films were dried at room temperature, 25°C. Eu^{3+} was incorporated into the titania thin films and used as a structural probe to study the influence of parameters such as deposition rate (100, 200, and 300 mm/min) and stirring (magnetic or ultrasound) employed during homogenization of the sol used for deposition.

The prepared films presented good transparency. Figures 22 and 23 correspond to the excitation and emission spectra of the Eu^{3+} ion in the titania film obtained at three different deposition rates, respectively.

For all the films, there was a large band with maximum at 309 nm in the excitation spectrum of Eu^{3+} at a fixed Eu^{3+} emission band (612 nm), irrespective of the deposition rate. This band can be attributed to CTB, which denotes covalente degree between Eu-ligand. The lower the energy of this band, the more covalent the Eu-ligand interaction [76]. All the films present same degree of Eu-ligand covalence, which might be due to the establishment of a Eu^{3+}-oxygen bond in the inorganic TiO_2 chain.

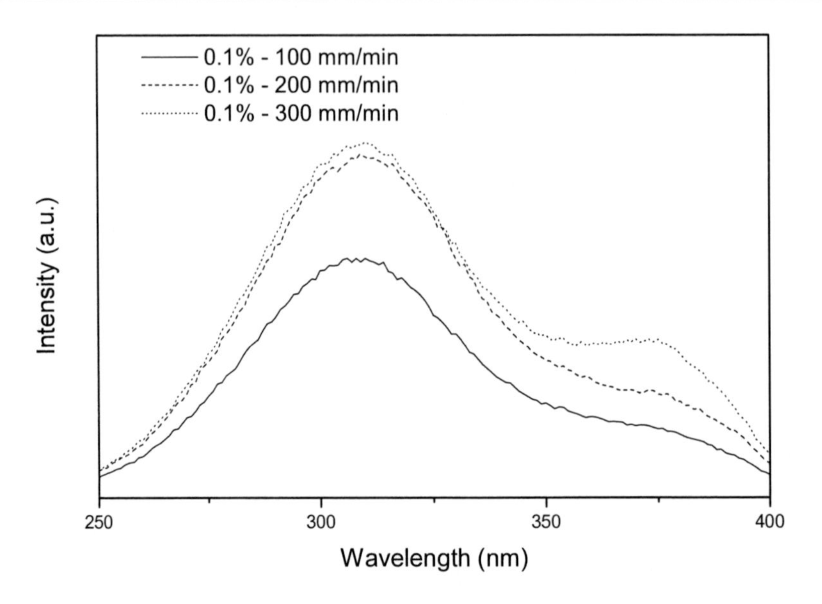

Figure 22. Excitation spectra of the Eu^{3+} ion in the films prepared at different deposition rates, the sol was stirred magnetically.

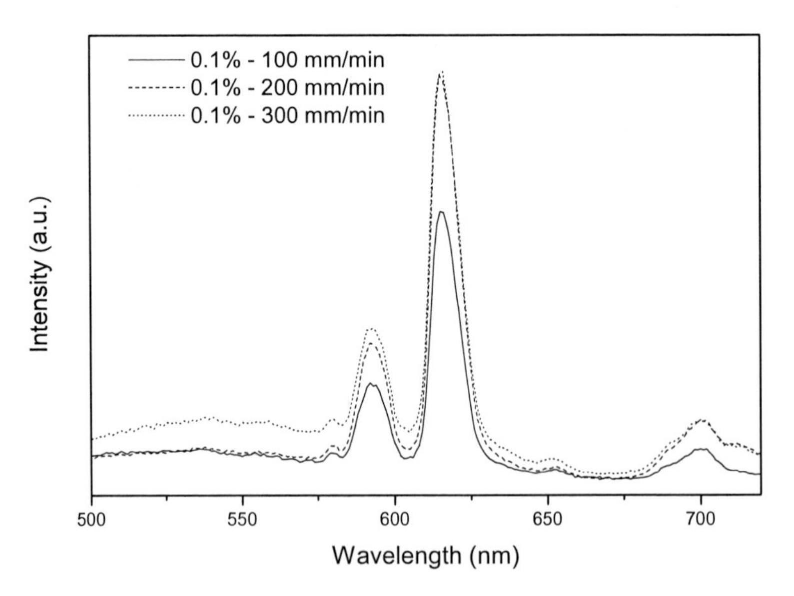

Figure 23. Emission spectra of the Eu^{3+} ion in the films prepared at different deposition rates, the sol was stirred magnetically.

Table 10. Relative area (RA) of the bands corresponding to the $^5D_0 \rightarrow {}^7F_0$ and $^5D_0 \rightarrow {}^7F_2$ transition with respect to the $^5D_0 \rightarrow {}^7F_1$ transition

Ultrasonic stirring	$^5D_0 \rightarrow {}^7F_0/{}^5D_0 \rightarrow {}^7F_1$	$^5D_0 \rightarrow {}^7F_2/{}^5D_0 \rightarrow {}^7F_1$
Speed 100 mm/min	0.22	1.94
Speed 200 mm/min	0.17	2.23
Speed 300 mm/min	0.14	2.62
Magnetic stirring		
Speed 100 mm/min	0.15	2.37
Speed 200 mm/min	0.18	2.24
Speed 300 mm/min	018	2.28

The emission spectra displayed the bands corresponding to the transitions arising from the Eu^{3+} excited state ($^5D_0$) to the fundamental ($^7F_{J\,=\,0,\,1,\,2,\,3\text{ and }4}$) level, when excited at 309 nm. The stirring mode used for film preparation (magnetic or ultrasonic) had no effect on the emission spectra obtained for the Eu^{3+} ion doped into the TiO_2 films, being the $^5D_0 \rightarrow {}^7F_2$ transition relatively more intense than $^5D_0 \rightarrow {}^7F_1$. The dependence of the emission spectra on the surrounding of the Eu^{3+} ion can be examined by the relative intensities, measured in terms of Relative Area (RA). Table 10 depicts the RA ratios of the bands $^5D_0 \rightarrow {}^7F_0$ and $^5D_0 \rightarrow {}^7F_2$ as a function of $^5D_0 \rightarrow {}^7F_1$.

For the samples homogenized by ultrasonic stirring, table 10 gives evidence of an enhancement in the relative intensity of the bands. This behavior may be due to better molecular mixing by this type of stirring. As for the samples prepared by magnetic stirring, the Eu^{3+} ions have similar surroundings.

The refractive index and thickness as a function of the deposition rates were obtained for the studied waveguides and calculated at 632.8 nm, using the parameters furnished by *m*-line measurements. The film thickness depended on the deposition rate. There was a sharp rise in thickness when the deposition rate was increased from 100 to 200 mm/min (0.1562 to 0.4552 nm), but this rise was less pronounced when the rate was elevated from 200 to 300 mm/min (0.4552 to 0.5364 nm). The refractive indices of the samples obtained at deposition rates of 200 and 300 mm/min were 1.9798 and 1.9803, respectively, which differ from the index achieved for the film deposited at 100 mm/min (2.3635).

The behavior of Eu^{3+} ions doped into thin TiO_2 films as a function of temperature was investigated. Ti sols were prepared from TEOT stabilized with acac at a 1:1 molar ratio. An ethanolic solution of $EuCl_3$ was prepared from the respective oxide. The sol was prepared with ethanol (HPLC degree) and acac, homogenized by magnetic stirring, followed by slow addition of the TEOT precursor and 15-minute stirring. $EuCl_3$ was added to the sol at a molar percentage of 0.3%, resulting in a yellow transparent sol. The films were obtained by the dip-coating technique. Glass substrates were carefully cleaned, immersed in the sols, and then dried at 100, 200, 300, 400, or 500°C for 2 hours.

Figure 24 presents the excitation spectra of Eu^{3+} in titania films treated at different temperatures.

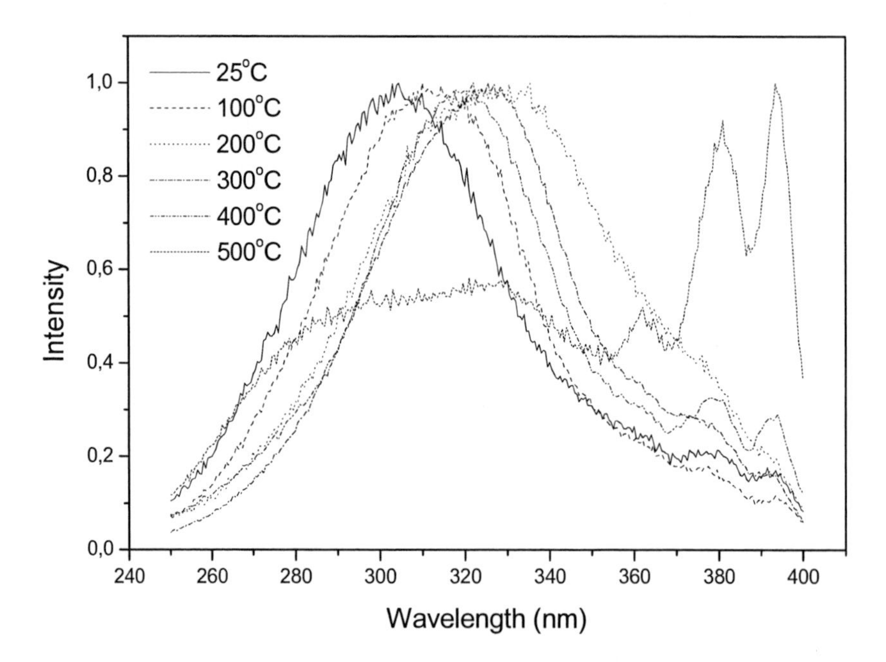

Figure 24. Excitation spectra of Eu^{3+} ions in titania films heat-treated at different temperatures.

Table 11. Excitation maxima for titania films treated at different temperatures

T (°C)	25	100	200	300	400	500
$\lambda_{exc.}$(nm)	303	310	320	325	320	393

The excitation maxima were dependent on heat-treatment temperature. When the films were heated over 300°C, the excitation bands corresponding to the Eu^{3+} ions appeared at 394 nm. Table 11 shows the excitation maxima in the films; the large band occurring in the 300-325 nm region was ascribed to a CTB.

The films treated at temperatures of up to 300°C displayed larger degree of covalence between the Eu^{3+} ions and the ligand, probably due to formation of a Eu-oxygen bond in the inorganic TiO_2 chain. At 500°C, the presence of an $^5L_6$ transition indicated the migration of the Eu^{3+} ions away from the inorganic chain.

Figure 25 depicts the emission spectra of Eu^{3+} in the films treated at different temperatures.

Excitation at 394 nm was used to raise the Eu^{3+} ions to the $^5L_6$ level. Figure 25 shows the lines representing transitions from the $^5D_0$ level to the different ^7F_J levels. The bandwidth was due to the non-homogeneous occupation of sites in the films [77].

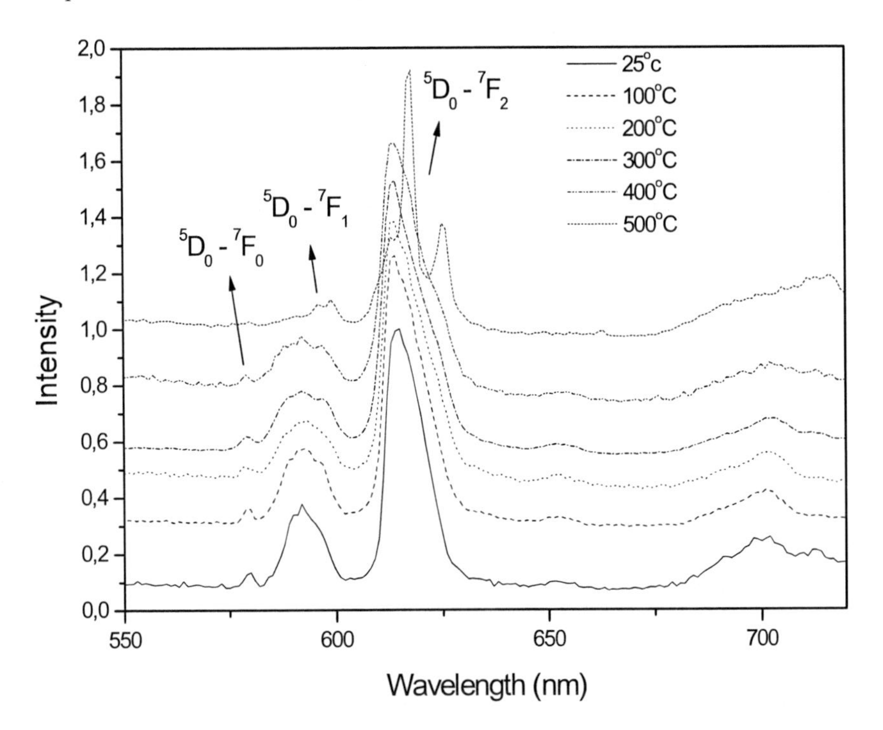

Figure 25. Emission spectra of Eu^{3+} ions in titania films treated at different temperatures.

The relative intensity of the $^5D_0 \rightarrow {}^7F_2$ transition with respect to the $^5D_0 \rightarrow {}^7F_1$ transition increased, which was consistent with the low symmetry surrounding the Eu^{3+} ions [78], and was due to gradual loss of water molecules as the temperature was raised to 400°C. The ratio was substantially enhanced at temperatures above 400°C, probably due to phase transition.

Table 12 displays the relative intensity of the $^5D_0 \rightarrow {}^7F_0$ and $^5D_0 \rightarrow {}^7F_2$ transitions with respect to $^5D_0 \rightarrow {}^7F_1$. The gradual elevation in the ratio between transitions indicates a change in the surroundings of the Eu^{3+} ion.

Table 12. Relative intensity ratio of the $^5D_0 \rightarrow {}^7F_0$ / $^5D_0 \rightarrow {}^7F_1$ and $^5D_0 \rightarrow {}^7F_2$ / $^5D_0 \rightarrow {}^7F_1$ transitions

samples	25oC	100oC	200oC	300oC	400oC	500oC
5D0 → 7F0/5D0 → 7F1	0.25	0.25	0.25	0.23	0.27	0.38
5D0 → 7F2/5D0 → 7F1	1.35	1.65	1.84	2.03	2.05	3.77

The lifetime of the emission at 613 nm and excitation at 394 nm evidenced an increase in lifetime with temperature, up to 400°C. This is an indication of loss of coordination water molecules around the Eu^{3+} ions. The Eu^{3+} emission lifetime in the case of the film heated at 500°C was reduced. This is probably due to phase transition, which promoted migration of the ions to the external surface, as observed by transmission electron microscopy [17].

The spectra in Figure 25 indicate the surroundings of Eu^{3+} ions are the same in the samples heat-treated at 25, 100, 200, 300, and 400°C; the wide band may indicate an amorphous TiO_2 structure [79]. The presence of the $^5D_0 \rightarrow {}^7F_0$ transition reveals that the local Eu^{3+} site is noncentrosymmetrical. The sample heated at 500°C displayed a different emission spectrum due to a change in the symmetry of the local site of Eu^{3+}, with the $^5D_0 \rightarrow {}^7F_2$ transition presenting a major definition, which demonstrates a phase transition in the inorganic chain composed of TiO_2.

The thickness of the films was qualitatively estimated by UV-Vis spectroscopy. The number of interference bands in the UV-Vis electronic spectrum may be related to the thickness. In reference [18], the thickness obtained for the films by m-line apparatus was 455.2 nm, which, according to UV-Vis spectroscopy, corresponds to four interference bands. Qualitatively, one can estimate ~400 nm for the sample dried at 25°C, and two interference bands for the samples treated at 100 and 200°C, indicating a thickness of ~200 nm. Lastly, the samples treated at 300, 400, and 500°C presented only one interference band, to which a thickness of ~100 nm was ascribed. The number of interference bands in the film treated at 25°C suggested that this film was thicker than the ones treated at other temperatures. The difference in thickness was attributed to loss of solvent molecules.

Films prepared with titanium and silicon alkoxide were examined. The Ti-Si sols were prepared by using TEOS and TEOT, acac, and $EuCl_3$ ethanolic solution. The utilized molar ratio was 1: 1: x: 0.1 for TEOS/TEOT/acac/ $EuCl_3$, respectively, where x = 1, 2, 3, or 4, and corresponding to a molar concentration in mol/L of 0.1:0.1:x:0.01, where x = 0.1, 0.2, 0.3, or 0.4 mol/L. The TEOS and TEOT were dissolved in dry ethanol containing acac. The resulting solutions were homogenized by ultrasonic stirring carried out for 30 min. Film deposition on glass substrate was accomplished by spin- or dip-coating.

The films were prepared with different amounts of the chelating agent (acac). Table 13 presents the molar ratio of the sols.

Table 14 depicts the thickness of the films prepared by the spin- and dip-coating techniques. The films obtained by spin-coating were thicker than those prepared by dip-coating.

The films exhibited good transparency in the 500-2700 nm range. The transparency of the films obtained by spin and dip-coating was 80 and 90%, respectively, which was ascribed to thickness.

Figure 26 corresponds to the emission spectra of Eu^{3+} ions doped into SiO_2-TiO_2 films containing different acac/Ti molar ratios, prepared by either spin-coating or dip-coating.

The emission spectra are similar, thus indicating that the Eu^{3+} ion occupies similar sites in the films. In other words, the preparation technique had no effect on the Eu^{3+} surroundings, only thickness was affected. When the emission spectra of Figure 26 are compared with those obtained for other systems containing Eu^{3+} and Ti (Figure 23 and 25), their similarities are easily detected. This is an indication that Eu^{3+} could be dispersed in the titanium matrix. The scheme 7 presents schematic film structure.

In this part of the work we prepared films containing the Eu^{3+} complex by the hydrolytic sol-gel methodology. Eu^{3+} ions were incorporated into these films, and different molar ratios of the ligand were employed. All the films were prepared by the dip-coating technique.

Table 13. Molar ratio of the sols

sample	Si	Ti	Acac	Eu3+	H2O	EtOH
1	1.0	1.0	0.5	0.1	0.7	4.5
2	1.0	1.0	1.0	0.1	0.7	4.5
3	1.0	1.0	1.5	0.1	0.7	4.5
4	1.0	1.0	2.0	0.1	0.7	4.5

Table 14. Thickness (μm) of the films obtained by spin and dip-coating

Samples	Spin-coating	Dip-coating
1	-	0.39 μm
2	1.0-1.5μm	0.40 μm
3	1.1-1.3μm	0.40 μm
4	1.0-1.2μm	0.21 μm

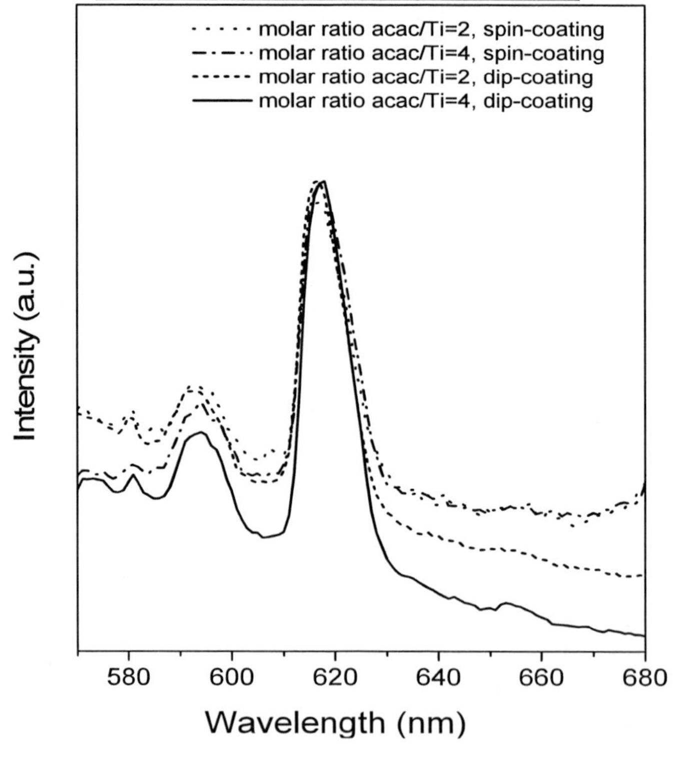

Figure 26. Emission spectra of the Eu^{3+} ion doped into SiO_2-TiO_2 films obtained by either spin-coating or dip-coating.

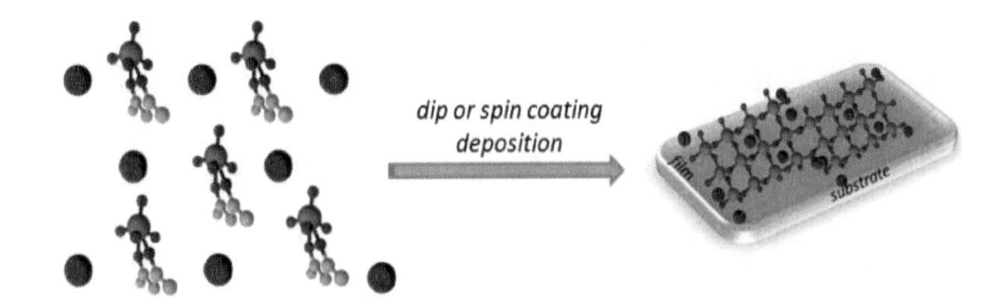

Scheme 7. Schematic representation of the titania-silica film obtained by sol-gel.

The sol was prepared by reaction of the dibenzoylmethane (DBM) sodium salt with the alkoxide 3-chloropropyltrimethoxysilane (ClPTMS) for 24 h, at 50°C [80]. TEOS was added after this period. After another 15 min, $EuCl_3$ methanolic solution was also added. The sol was separated into three parts. The DBM ligand was added at an ion/ligand molar ratio of 1:1, 1:3, or 1:10. The sol was deposited onto a glass substrate by the dip-coating technique.

Evidence of the presence of the DBM complex in the film is given in Figure 27, which shows the absorption spectra of the ligand DBM and the film.

The large band with maximum at 342 nm (Figure 27a) is ascribed to the absorption of DBM-Na. A similar band with absorption maximum at 347 nm is observed for the film containing the Eu^{3+}-DBM compounds, strongly indicating that DBM is present in it.

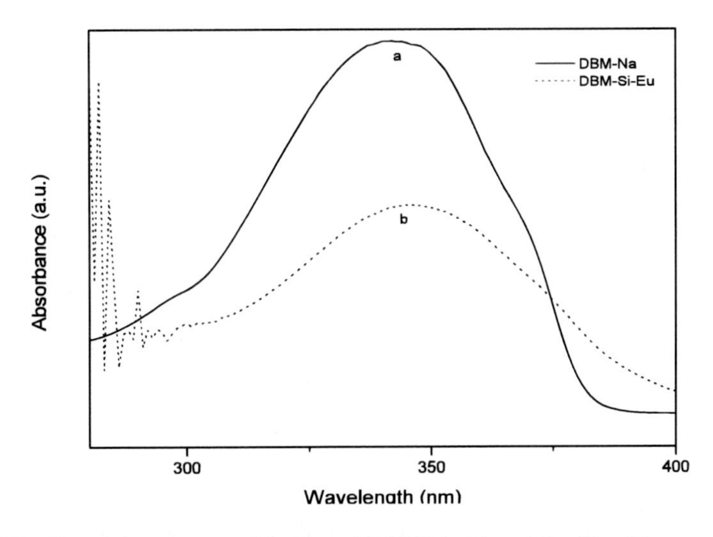

Figure 27. Absorption spectra of the ligand DBM-Na (a) and the film (b).

Figure 28 shows the excitation spectra of the Eu^{3+}-DBM compounds in the films containing different ligand molar ratios, namely 1:1, 1:3, and 1:10.

In this case there was a reduction in the relative intensity of the excitation band, ascribed to intramolecular process of ligand-metal energy transfer.

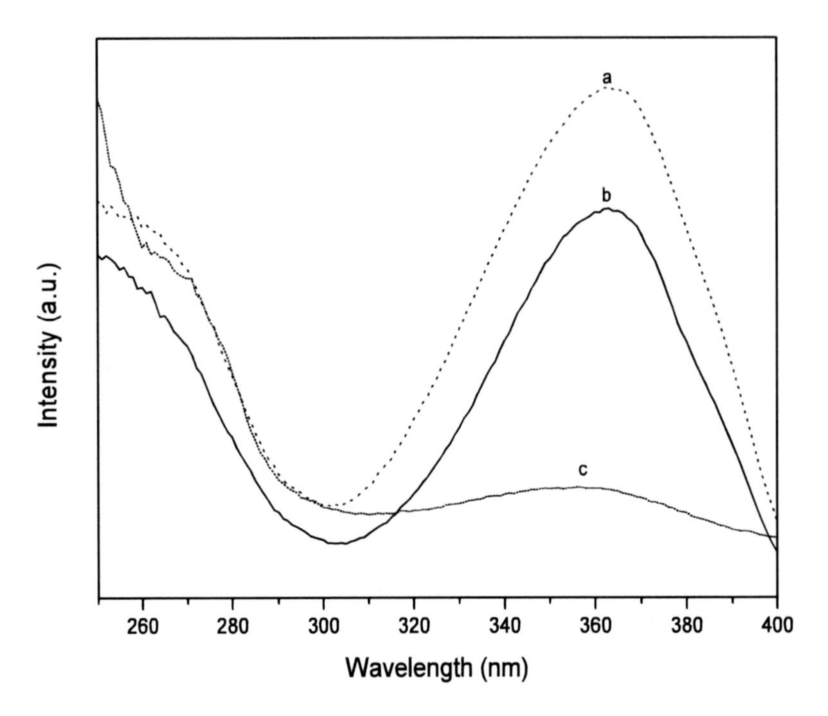

Figure 28. Excitation spectra of the Eu^{3+}-DBM compounds films containing different ion/ligand molar ratios (a) 1:1, (b) 1:3, and (c) 1:10. $\lambda_{em} = 612$ nm.

Figure 29 illustrates the emission spectra of the Eu^{3+} ion in the films containing the Eu^{3+}-DBM compound.

The emission spectra of the DBM-Si-Eu fims prepared with ion/ligand molar ratios of 1:1 and 1:10 have a similar profile, whilst the sample containing an ion/ligand molar ratio of 1:3 behaves differently. In the latter case, the relative intensity of the $^5D_0 \rightarrow {}^7F_0$ and $^5D_0 \rightarrow {}^7F_2$ transitions is more pronounced than in the spectra of the former samples, not to mention the fact that the $^5D_0 \rightarrow {}^7F_2$ transition is split, indicating that the Eu^{3+}-DBM compound is present in the film. The relative intensity of the $^5D_0 \rightarrow {}^7F_0$ and $^5D_0 \rightarrow {}^7F_2$ transitions with respect to the $^5D_0 \rightarrow {}^7F_1$ transition presents values around 0.11 and 4.40, evidencing that the Eu^{3+} ion occupies a low symmetry site.

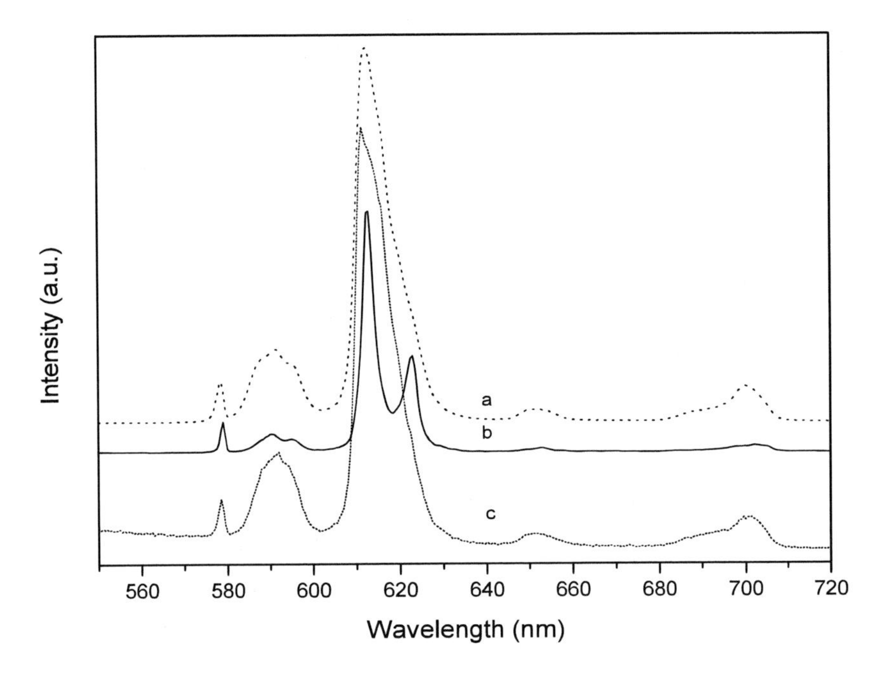

Figure 29. Emission spectra of the films containing the Eu^{3+}-DBM compounds, at different ion/ligand molar ratios, (a) 1:1, (b) 1:3, and (c) 1:10. λ_{em} = 350 nm.

Other kinds of hybrid materials make use of Eu^{3+} ion as structural probe, in order to amplify the emission properties and enable their use in several applications, such as optical fibres, electroluminescent materials, among others [81]. Organic ligands that absorb energy and transfer it to the ion (antenna effect) are employed in the preparation of the lanthanide complex, so as to promote intensification of the luminescence. The complex can be immobilized onto rigid matrices, to reduce energy loss by vibrational modes. In this sense, some interesting matrices are mineral clays, whose morphological characteristics allow for surface and interplanar chemical modifications that lead to effetive complex immobilization [82 - 85]. In the next part of the discussion, we will describe the functionalization of kaolinite from São Simão, Brazil, with pyridine-2-carboxylic (PIC) or pyridine-2-6-dicarboxylic (DIP) acids by displacement of dimethyl sulfoxide (DMSO) molecules from the previously formed kaolinite-DMSO complex. The methodology described by Detellier and co-workers was employed in the preparation of the precursor intercalated with (Ka-DMSO) [86, 87]. The hybrid organic-inorganic materials were obtained by keeping a mass of the precursor (Ka-DMSO) in the presence of the melted acids PIC or DIP for 40 h. The resulting solids were designated

KPIC and KDIP, respectively. The samples KPIC and KDIP were suspended in $EuCl_3$ ethanolic solution at different ion/ligand molar ratios, namely 1:1, 1:2, and 1:3.

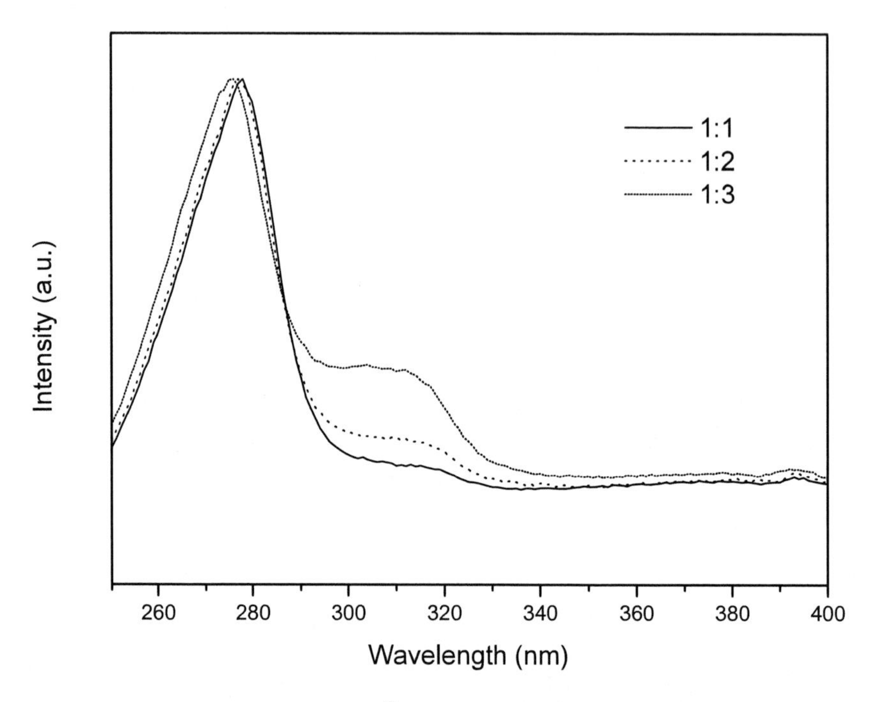

Figure 30. Excitation spectra of the Eu^{3+} ion in hybrid organic/ inorganic materials (ka-pa)Eu(1:1), (ka-pa)Eu(1:2), and (ka-pa)Eu(1:3), with different ion/ligand molar ratios, λ_{em} = 612 nm.

Table 16. Relative intensity of the $^5D_0 \rightarrow {}^7F_0$ / $^5D_0 \rightarrow {}^7F_1$ and $^5D_0 \rightarrow {}^7F_2$ / $^5D_0 \rightarrow {}^7F_1$ transitions

Sample	5D0→7F0/5D0→7F1	5D0→7F2/5D0→7F1
(Ka-pa)Eu (1:1)	0.04	1.83
(Ka-pa)Eu (1:2)	0.13	2.06
(Ka-pa)Eu (1:3)	0.16	2.11

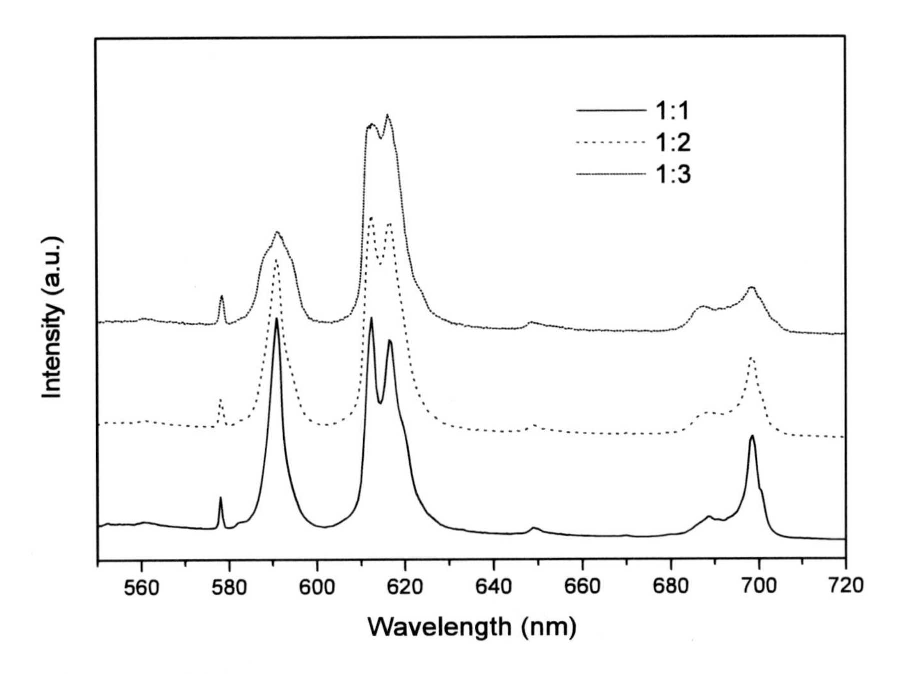

Figure 31. Emission spectra of the Eu^{3+} ion in hybrid organic/ inorganic materials (ka-pa)Eu(1:1), (ka-pa)Eu(1:2), and (ka-pa)Eu(1:3), with different ion/ligand molar ratios, $\lambda_{exc} = 277$ nm.

The excitation spectra of the Eu^{3+} incorporated into the kaolinite containing acids at different molar ratios are represented in Figure 30. The large band appears at a maximum of 277 nm, which is different from the maxima of the ligands and the Eu^{3+} ion. Therefore, this band was ascribed to ligand-metal charge transfer (LMCT) [88], as observed in the spectra described by Reinhard, Binnemans, Zolin, Lima, and others [89 - 91].

Figure 31 corresponds to the emission spectra of the Eu^{3+} ion in the samples containing different ion/ligand molar ratios. The typical transition of the Eu^{3+} ion, from $^5D_0$ to ^7F_J (J = 0, 1, 2, 3 and 4) is detected [92].

The emission spectra of the Eu^{3+} ion in the luminescent hybrid materials present only one band corresponding to the $^5D_0 \rightarrow {}^7F_0$ transition, for all the samples. This band becomes larger with increased ion/ligand molar ratio, revealing formation of a new symmetry site. The presence of this transition, with high relative intensity, indicates that the Eu^{3+} ion occupies a site of lower symmetry

The relative intensities of the $^5D_0 \rightarrow {}^7F_0$, $^5D_0 \rightarrow {}^7F_1$ and $^5D_0 \rightarrow {}^7F_2$ transitions of these hybrid materials are different compared with those of the complexes Eu-PIC and Eu-DIP, as observed by Reinhard et al [89]. The relative intensity of these bands are also different the when ion/ligand molar ratio is compared. Generally, the larger ratio between the $^5D_0 \rightarrow {}^7F_2$ / $^5D_0 \rightarrow {}^7F_1$ transitiion is evidence that the Eu^{3+} ion occupies a site of lower symmetry. In Table 16 the relative intensities of the $^5D_0 \rightarrow {}^7F_0$ and $^5D_0 \rightarrow {}^7F_2$ transitions with respect to the $^5D_0 \rightarrow {}^7F_1$ transition is shown. The gradual increase in the ratio between the transitions suggests change in the surroundings of the Eu^{3+} ion.

These results shows that elevation in the ion/ligand molar ratio causes an enhancement in the relative intensity of the $^5D_0 \rightarrow {}^7F_0/{}^5D_0 \rightarrow {}^7F_1$ and $^5D_0 \rightarrow {}^7F_2/{}^5D_0 \rightarrow {}^7F_1$ transitions, suggestive of reduced Eu^{3+} symmetry [93].

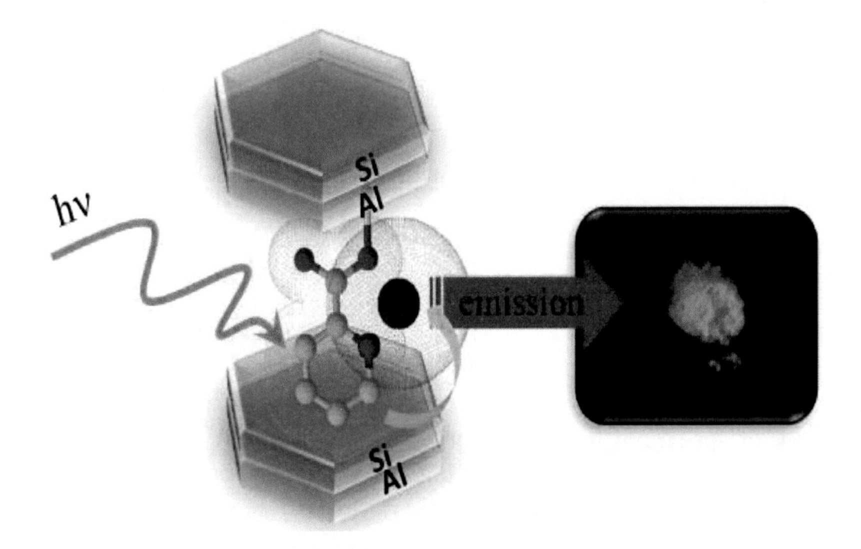

Scheme 8. Schematic antenne effect.structure of the kaolinite with Eu-complex.

The enhanced lifetime of the emission corresponding to the $^5D_0 \rightarrow {}^7F_2$ transitions with rising ion/ligand molar ratio is an indication that water molecules coordinated to the Eu^{3+} ion were replaced with the ligand.

During this entire investigation it was possible to clearly observe the influence of the Eu^{3+} surroundings on the emission spectra. The change in symmetry can be noticed from the relative intensity of the band in the emission spectra. The scheme 8 and 9 present schematic structure of the antenne effect and kaolinite with Eu-complex, respectively.

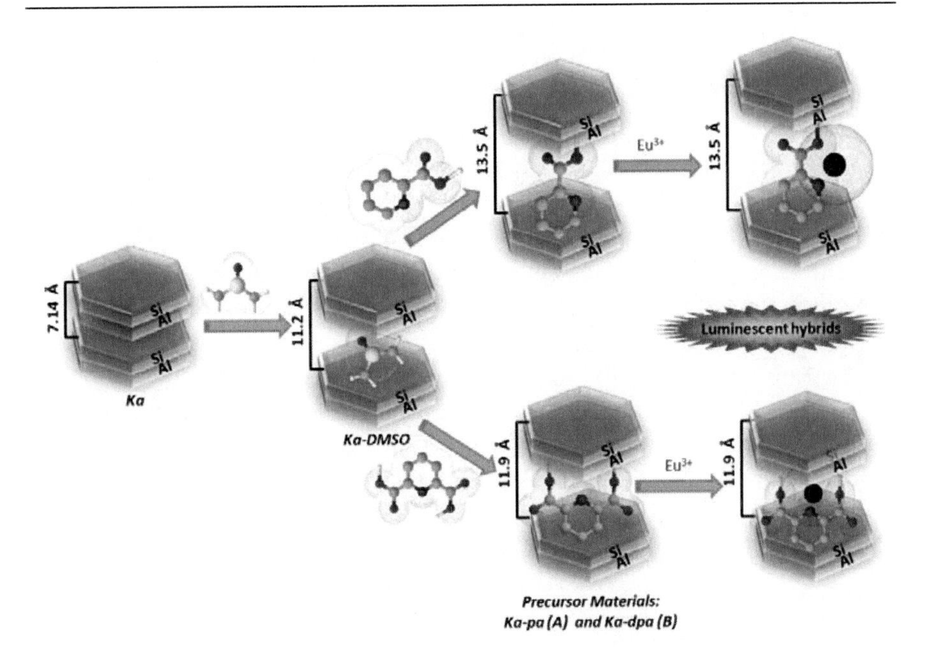

Scheme 9. Schematic structure of the kaolinite with Eu-complex.

NON-HYDROLYTIC SOL-GEL PROCESS

In the second section of this chapter we shall discuss the different emission spectra of the Eu^{3+} ion in matrices prepared by the non-hydrolytic sol-gel route. These matrices find several kinds of applications, such as biomaterials, glass ionomer, solid state lasers, phosphors, and others.

The materials based on silicon and aluminum were synthesized in three different compositions, using the nonhydrolytic sol-gel route, and were dubbed Glasses A2, A3.3, and A4. The compositions of Glasses A2 and A4 were based on the formulations proposed by Wilson and Hill [94, 95]. The material designated Glass A3.3 was synthesized using the Hill formulation, but with a higher phosphorus concentration. The P/Al ratio in the A2, A3.3, and A4 materials was 0.06, 0.45, and 0.33, respectively. $EuCl_3$ (1% of the total theoretical mass) was added as structural probe. The solvent was rotaevaporated, and the remaining material was oven-dried at 50°C for 7 days, resulting in a white, solid material. Part of this material was heat-treated at 600°C for 4 hours.

Rare earth ions can be used as probes in the elucidation of a variety of structures, enabling the investigation of materials on the basis of the spectroscopic properties of these ions. Eu^{3+} has been the most commonly employed and studied rare earth ion because its emission spectrum is easy to interpret. In order to monitor the glassy environment of the materials synthesized in this study, Eu^{3+} ions were added to the matrices [96, 97].

The samples dried at 50°C showed no excitation or emission spectra. This is because the presence of a large quantity of molecules from the solvent and by products of the reaction promotes energy loss through vibrational mechanisms, as indicated by the thermal analyses. Another possible factor for the suppression of luminescence is incomplete condensation; i.e., the number of–OH groups on the surface of the materials can promote this suppression. The vibration of –OH groups occurs at 3500 cm^{-1}, while the $^5D_0$ level of the Eu^{3+} ion appears at approximately 17000 cm^{-1}. Hence, five –OH groups are needed to cause the non-radiative decay of a photon.

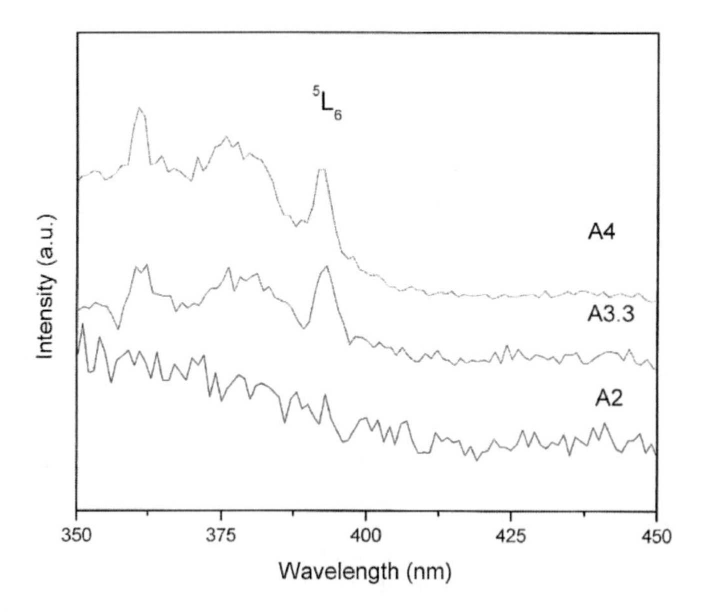

Figure 32. Excitation spectrum of the Eu^{3+} ion in the samples A2, A3.3, and A4 treated at 600°C and measured at 616 nm.

Sample A2, which was treated at 600°C (Figure 28), did not show excitation or emission spectra, probably because the structure of the material promoted non-radiative decay. Figure 32 presents the excitation spectra of the samples heat-treated at 600°C as measured at 616 nm. Samples A3.3 and A4 display the typical bands of the transition between the fundamental ($^7F_0$) and the excited ($^5L_6$) states of the Eu^{3+} ion.

Radiative emissions of the Eu^{3+} ion are observed at different wavelengths, each length being characteristic of a specific electronic transition. The wavelengths of approximately 580, 590, 610, and 700 nm correspond to the $^5D_0 \rightarrow {}^7F_0$, $^5D_0 \rightarrow {}^7F_1$, $^5D_0 \rightarrow {}^7F_2$, and $^5D_0 \rightarrow {}^7F_3$ transitions, respectively.

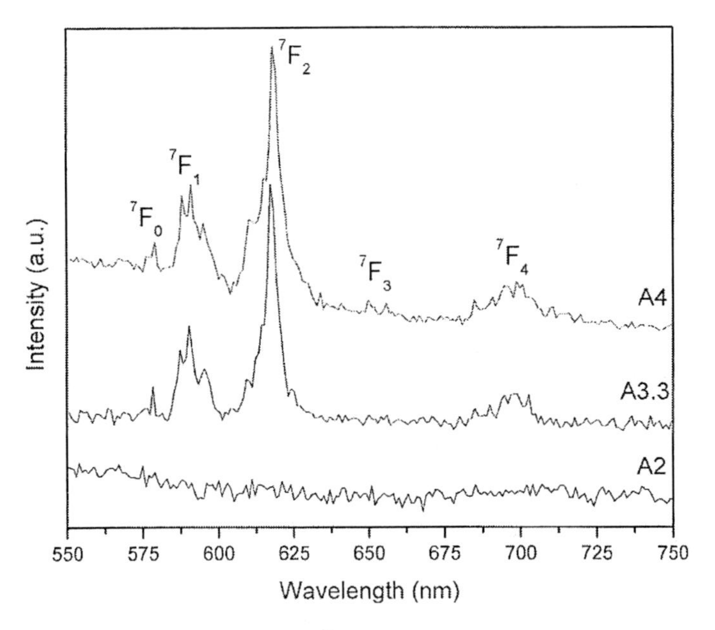

Figure 33. Emission spectrum of the Eu^{3+} ion in the samples A2, A3.3, and A4, treated at 600°C and excited at 392 nm.

Figure 33 depicts the emission spectra of the samples A2, A3.3, and A4 treated at 600°C and excited at 392 nm ($^5L_6$ level). Sample A2 has no ion emission. Samples A3.3 and A4 display a band corresponding to the $^5D_0 \rightarrow {}^7F_0$ transition, which indicates that the ions occupy sites without inversion centers. Also, the magnetic dipole $^5D_0 \rightarrow {}^7F_1$ and electric dipole $^5D_0 \rightarrow {}^7F_2$ transitions present more bands than those permitted by the unfolding of the crystalline field (2J+1), evidencing the presence of Eu^{3+} ion in different sites of the matrix. The width of the emission bands also demonstrates inhomogeneity of the Eu^{3+} sites. Thus, one can state that the chemical medium of the ion resembles that of a system presenting amorphous and/or mixed amorphous and crystalline structures.

The emission spectra for the samples obtained by the hydrolytic process presented a different profile (figures 25, 29 and 31). The large band in those spectra indicated that the Eu^{3+} ion occupied non-homogeneous sites (amorphous phases) only. The non-hydrolytic sol-gel process, on the other hand, produces crystalline phases with fine lines in the emission spectra.

The X-ray diffractograms of the A2 samples show the predominance of an amorphous structure. The A3.3 material displays an amorphous structure with crystalline phases attributed to fluorapatite ($Ca_5(PO_4)_3OH$) and mullite ($3Al_2O_32SiO_2$), according to Gorman et al. Sample A4 also presents an amorphous phase and a crystalline phase, which is ascribed to mullite [98].

In this study we also prepared matrices by the non-hydrolytic sol-gel route, to obtain a stable inorganic oxide for use as phosphor. Specifically, two different matrices were prepared, the first one containing Al-Gd-Ca and the second Al-Gd. The preparation of the gel was carried out in oven-dried glassware. $AlCl_3$ and ethanol were reacted with (or without) $CaCl_2$, as well as $EuCl_3$ and $GdCl_3$. The structural probe (Eu^{3+}) was added in a molar ratio of 1% in relation to the Gd^{3+} ion. The mixture was refluxed for 3 h at 110 °C, under argon atmosphere. The condenser was placed in a thermostatic bath at –5 °C. After 120 min, a solid material was obtained. After the reflux, the mixture was cooled and aged overnight in the mother liquor at room temperature (RT), to allow precipitation through aging in the mother liquor. The solvent was then removed under vacuum. The powders were dried and received heat treatment at 100, 400, 600, 800, and 1000 °C for 4 hours.

Figures 34 and 35 illustrate the excitation spectra of Eu^{3+}-doped Gd-Ca-Al and Gd-Al matrices, respectively, monitored at 616 nm ($^5D_0 \rightarrow {}^7F_2$).

The sharp lines are assigned to transitions from the $^7F_0$ level to the $^5L_6$ (394 nm), $^5L_7$ (382 nm), and $^5D_4$ (360 nm) levels in the samples treated at 600, 800, and 1000 °C. These lines are characteristic of the f $\leftrightarrow$ f transition of the rare earth ion. *The CTB* appears in the high-energy region; that is, at short wavelengths (between 250 and 300 nm). The CTB is a measure of the covalent character of the Eu-Ligand bond. The lower the transition energy, the stronger the interactions between the metal and the ligand. The maxima corresponding to the CTB (Figure 34) obtained for the samples treated at 800 and 1000°C appear in the same region (270 nm). As for the sample treated at 600 °C, the CTB is located at 280 nm. The bands shift to shorter wavelengths for the samples treated at lower temperatures, indicating that the Eu^{3+}-O^{2-} bond has different covalent character or different coordination sites.

In Figure 35, the CTB is located at 290 nm for the samples treated at 400 and 600 °C, whereas the CTB of the materials treated at 800° and 1000°C is located at 265 nm.

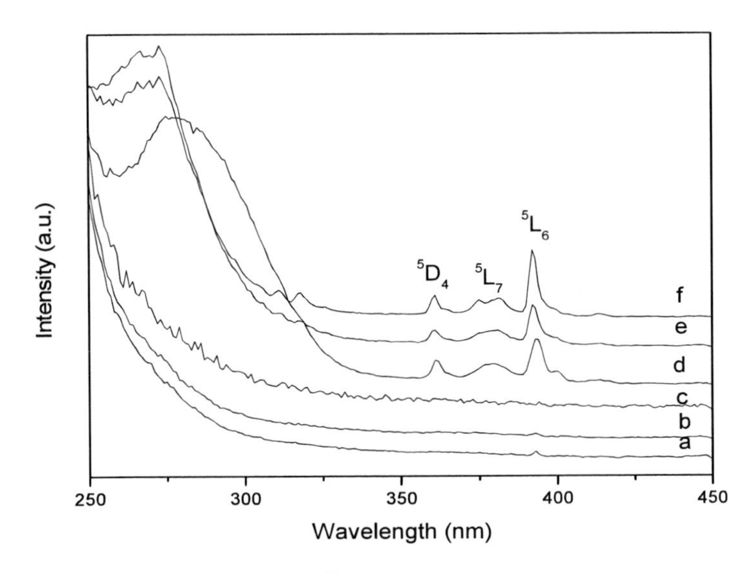

Figure 34. Excitation spectra of the Eu^{3+} ion in Gd-Ca-Al matrices treated at different temperatures, λ_{em} = 616 nm, (a) 50 °C , (b) 100 °C, (c) 400 °C, (d) 600 °C, (e) 800 °C, and (f) 1000 °C.

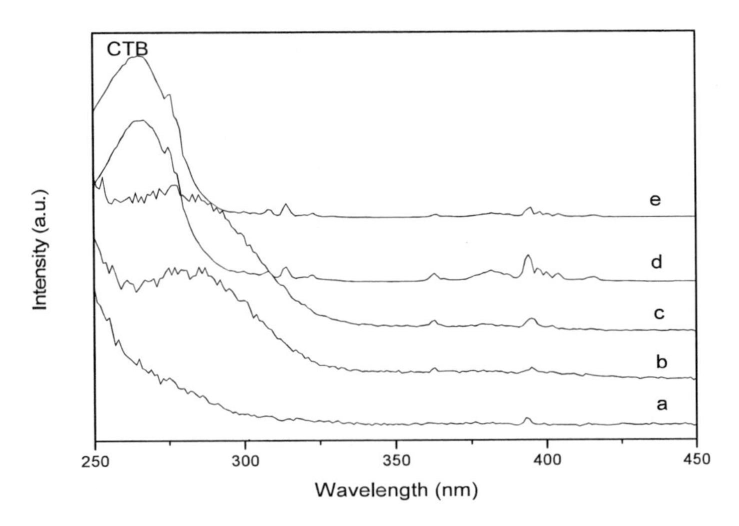

Figure 35. Excitation spectra of the Eu^{3+} ion in the Gd(Eu)-Al-O host treated at (a) 100, (b) 400, (c) 600, (d) 800 , and (e) 1000 °C for 4 hours, λ_{em} = 616 nm.

Figures 36, 37, and 38 display the emission spectra of the Eu^{3+} ion in the Gd-Ca-Al matrices, monitored at CTB, 391, and 463 nm , respectively.

The emission spectra of Eu^{3+} display bands at 578, 587, 595, 610, 650, and 695 nm, correponding to the electronic transition from the $^5D_0$ to the ^7F_J (J = 0, 1, 2, 3, and 4) level when the maxima is fixed at 275 and 463 nm, for the samples treated at 600, 800, and 1000 °C for 4 h. The large magnitude of the spin-orbit coupling in lanthanides causes the individual J levels of the various electronic terms to be well separated from each other, except for the ground $^7F_0$ and emissive $^5D_0$ states of Eu^{3+}, which are non-degenerated. The highly forbidden $^7F_0 \rightarrow {}^5D_0$ transition of Eu^{3+} is particularly important in that only a single transition is possible for a single Eu^{3+} ion environment [3]. The excitation at 391 nm ($^5L_6$ level) leads to emission for all the samples. The band corresponding to the $^5D_0 \longrightarrow {}^7F_2$ transition is more intense than that due to $^5D_0 \rightarrow {}^7F_1$. This indicates that the Eu^{3+} ion occupies a site without an inversion center, which can be due to $GdCaAl_3O_7$ phases.\

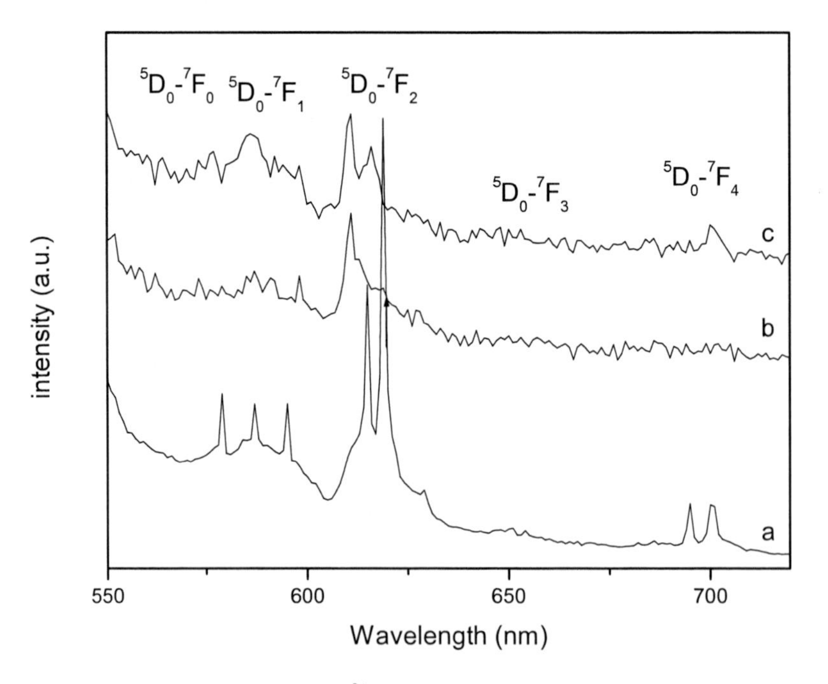

Figure 36. Emission spectra of the Eu^{3+} ion in the Gd-Ca-Al samples treated at (a) 600°C, (b) 800 °C, and (c) 1000 °C, excited at CTB.

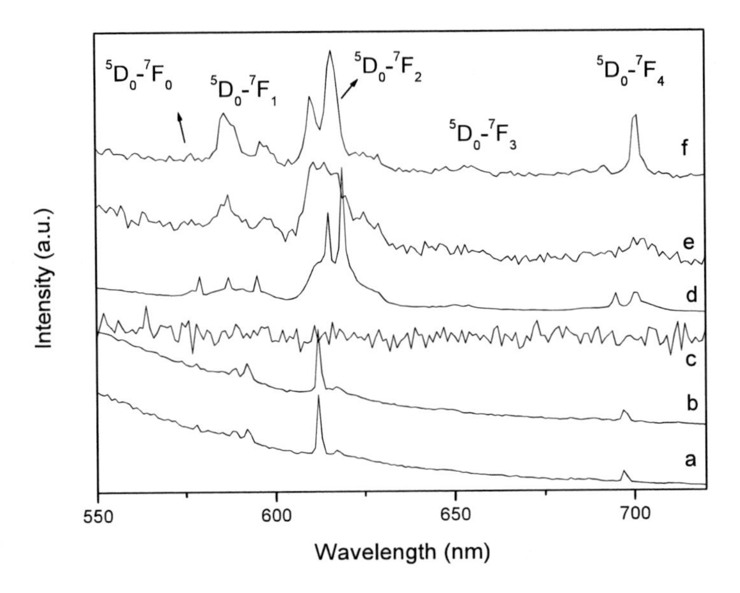

Figure 37. Emission spectra of the Eu^{3+} ion in the Gd-Ca-Al samples treated at (a) 50 °C, (b) 100 °C, (c) 400 °C, (d) 600 °C, (e) 800 °C, and (f) 1000 °C, excited at 391 nm ($^5L_6$ level).

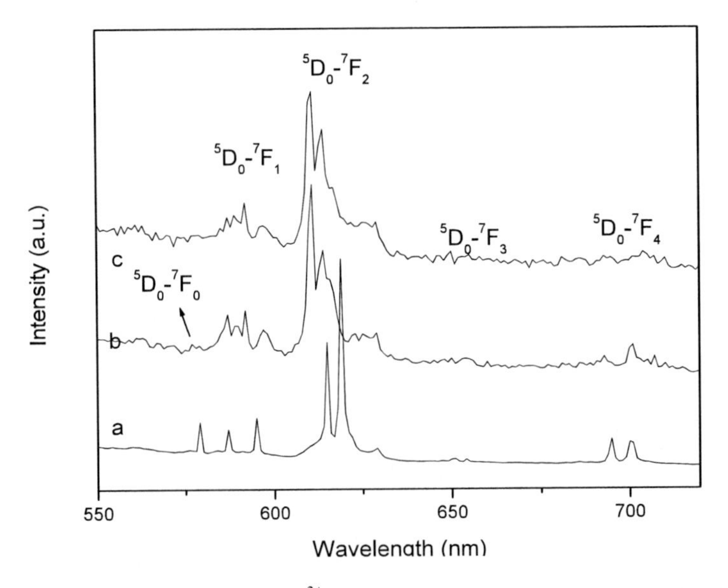

Figure 38. Emission spectra of the Eu^{3+} ion in the Gd-Ca-Al samples treated at (a) 600 °C, (b) 800 °C, and (c) 1000 °C, excited at 463 nm ($^5D_2$ level).

The emission spectrum of the sample treated at 600°C shows similar fine line emission, independent of the maximum excitation. This is an indication that there is probably one phase at this temperature. One band is due to the 0 → 0 transition, two are due to the 0 → 1 transition, and three are due to the 0 → 2 transition, which can be another indication of the presence of one phase in the $GdCaAl_3O_7$ matrix. The various J levels are further split by ligand fields in the maximum of 2J + 1; therefore, in the $^7F_0 \rightarrow {}^5D_1$ and $^7F_0 \rightarrow {}^5D_2$ transitions, the band numbers may be 3 and 5, respectively.

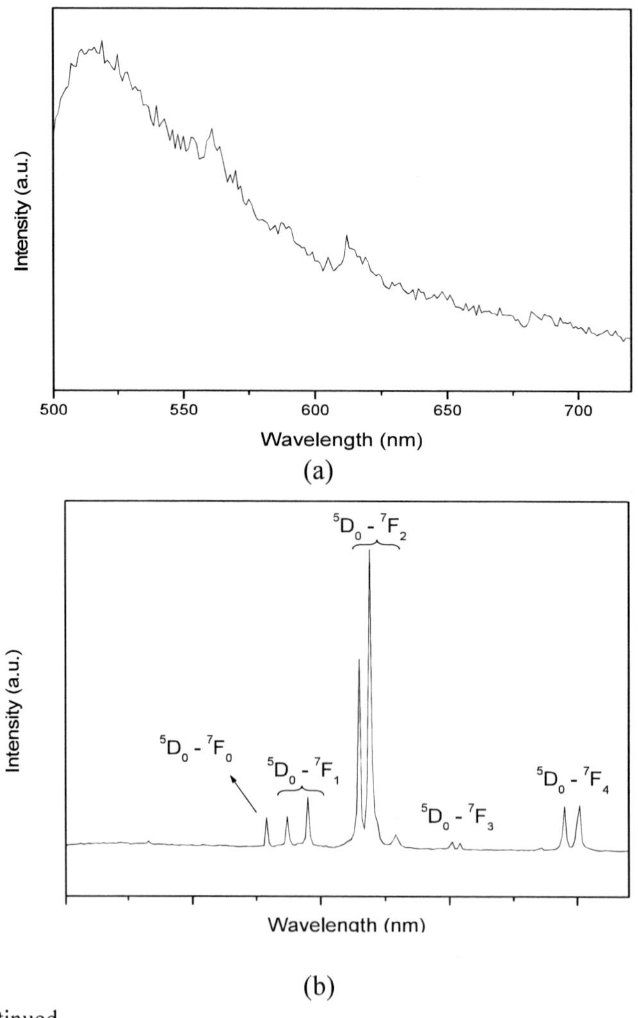

Figure continued

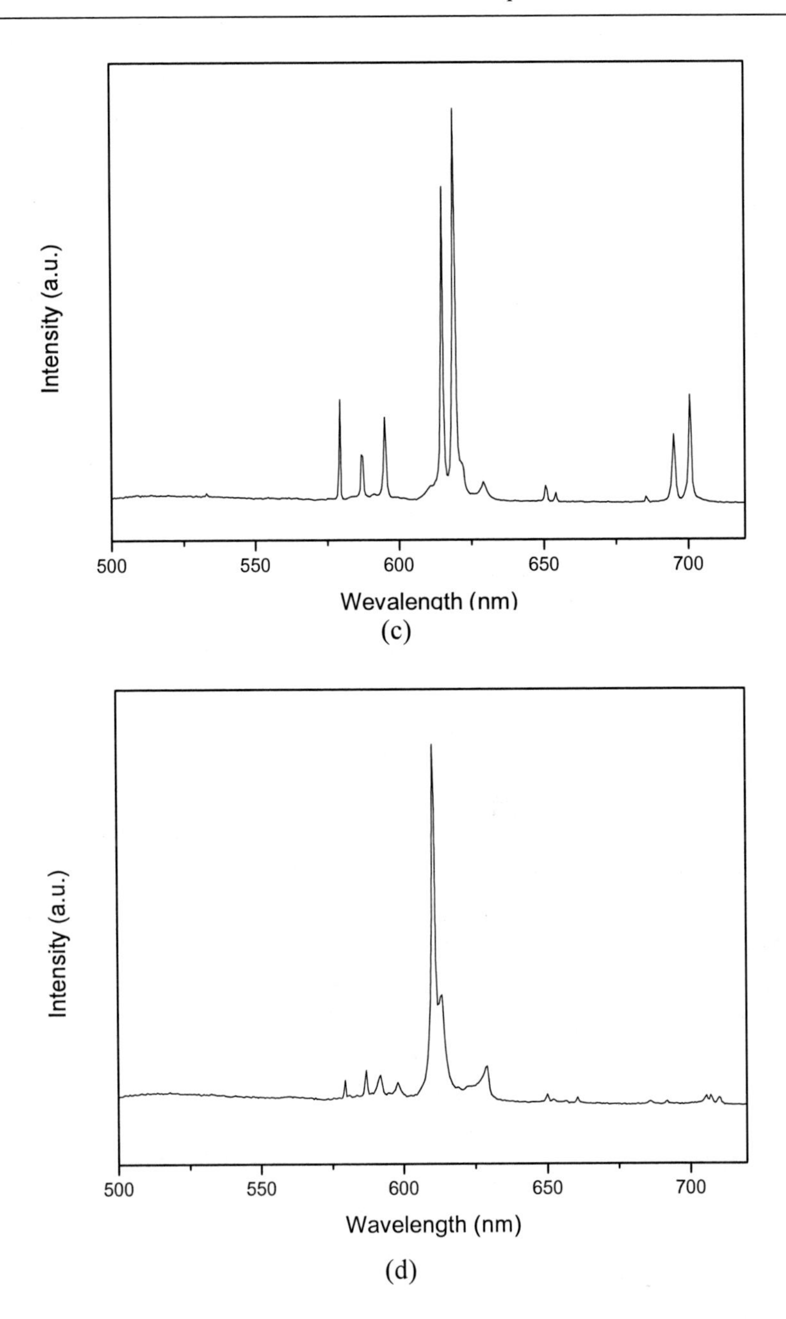

(c)

(d)

Figure continued

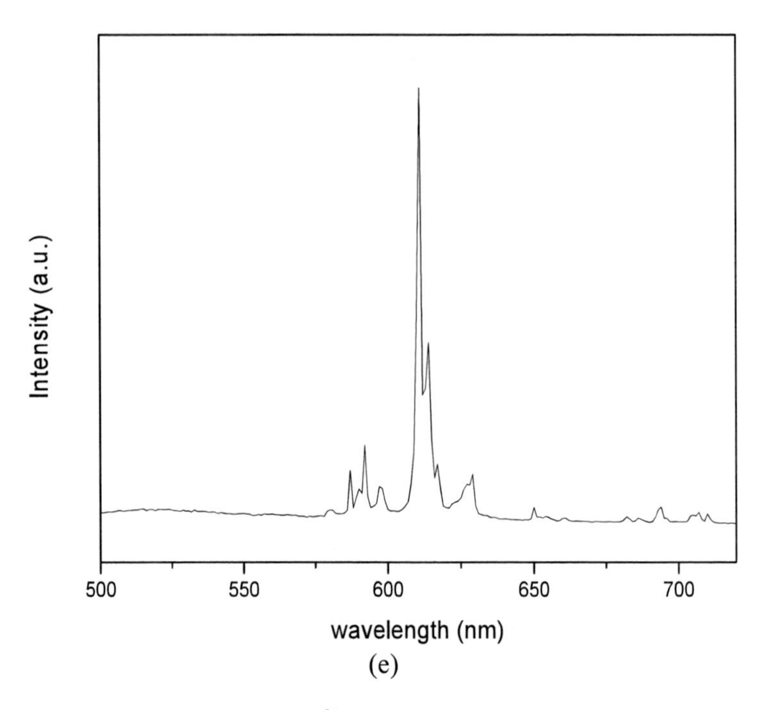

Figure 39. Emission spectra of the Eu^{3+} ion in the Gd(Eu)-Al-O samples treated at (a) 100, (b) 400, (c) 600, (d) 800, and (e) 1000°C for 4 hours, excited at the CTB.

The DTA and DSC curves reveal an endothermic transition between 100 and 200 °C, ascribed to loss of water and solvent molecules. The exothermic peaks at 400 and 960 °C are due to phase transition, which is also observed in the emission spectra and XRD patterns.

Figures 39, 40, and 41 correspond to the emission spectra of the Eu^{3+} ion in the Gd(Eu)-Al-O host, monitored at the CTB, 393 nm ($^5L_6$ level), and 467 nm ($^5D_2$ level), respectively.

The emission spectra of the phosphors, obtained by excitation at CTB, 393, and 467 nm, consist of the $^5D_0 \rightarrow {}^7F_J$ (J = 0 - 4) emission lines of Eu^{3+} dominated by the $^5D_0 \rightarrow {}^7F_2$ (~ 610 nm) electric dipole transition, which is strongly dependent on the Eu^{3+} surroundings.

Emission of Eu^{3+}-doped samples obtained at 100 °C present broad peaks or no peaks at all, typical of the emission profile of non-crystalline compounds or highly disordered systems (Figures 39 and 41). In Figure 40, the emission spectrum can be ascribed to halide precursors.

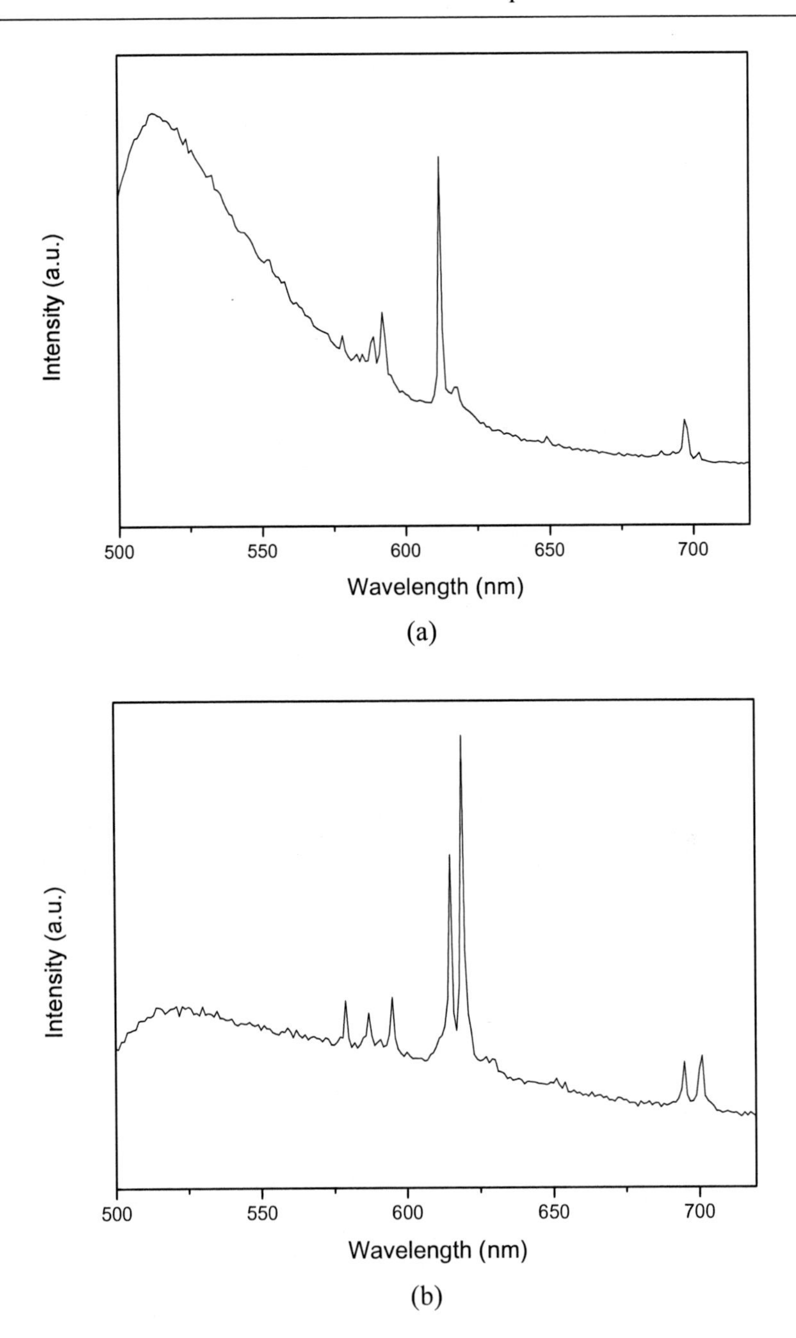

Figure continued

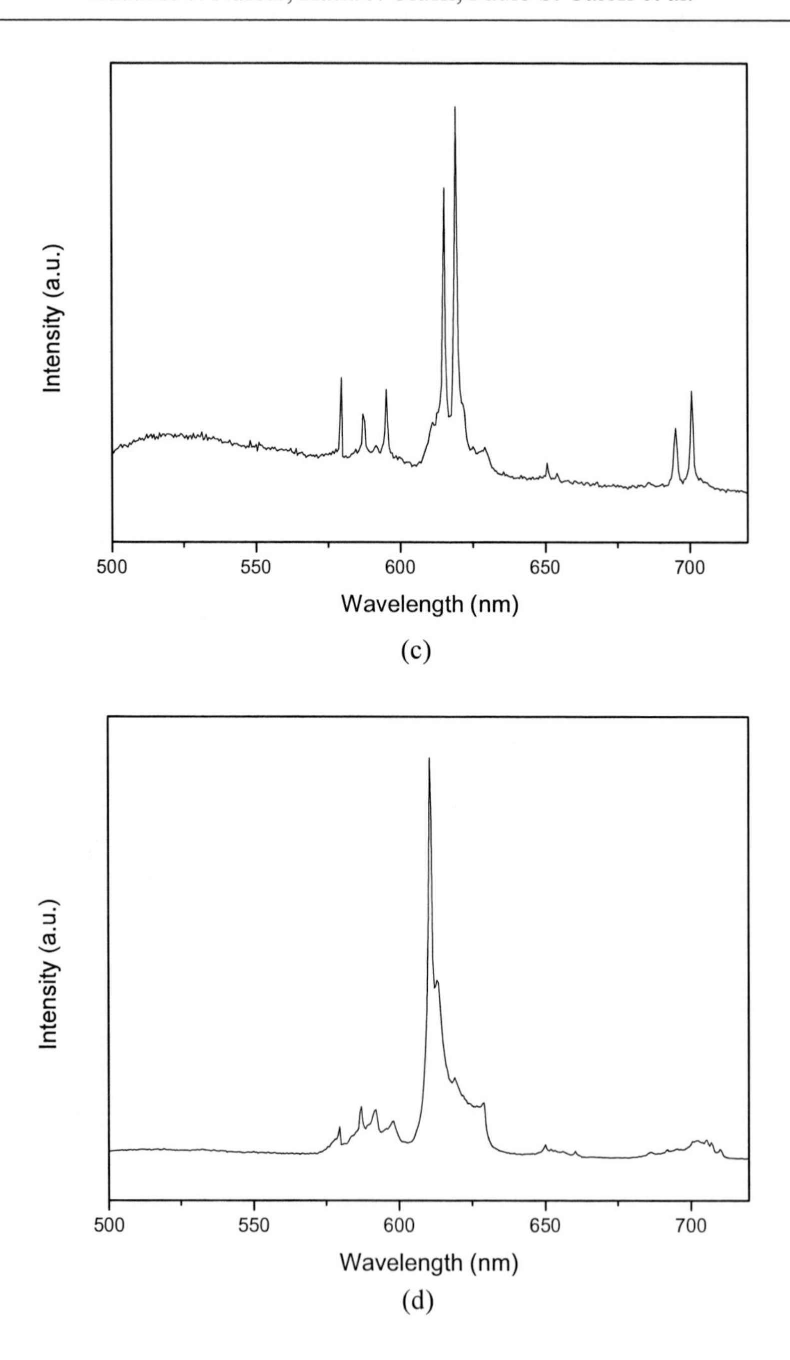

Figure continued

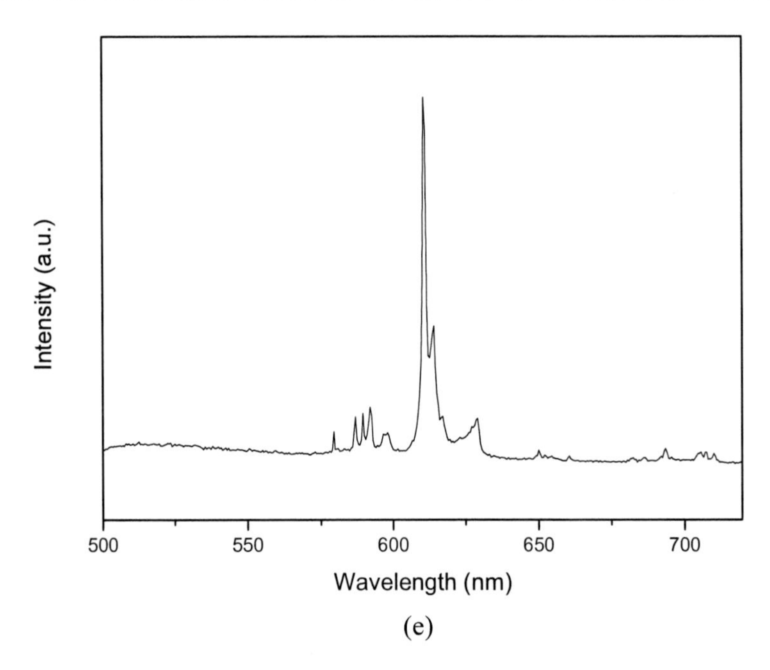

(e)

Figure 40. Emission spectra of the Eu^{3+} ion in the Gd(Eu)-Al-O samples treated at (a) 100, (b) 400, (c) 600, (d) 800, and (e) 1000 °C for 4 hours, excited at 393 nm ($^5L_6$ level).

A clear change in the symmetry of Eu^{3+} can be noted, since the samples present different numbers of emission lines. This indicates different symmetries for the Eu^{3+} ion.

The $^5D_0 \rightarrow {}^7F_0$ transition appears at the same wavelength for all the samples, independent of the excitation wavelength. There is a difference concerning the relative intensity with relation to the $^5D_0 \rightarrow {}^7F_1$ transition. The latter transition presents different numbers of bands, depending on the excitation wavelength and temperature. The samples treated at 400 and 600°C display a similar number of bands, thus indicating that Eu^{3+} occupies the same symmetry site in the host, as observed by XRD. For the powders heated at 800 and 1000°C, new profile emission spectra are observed. This is due to a new symmetry site for Eu^{3+}, also seen in the XRD. The sample treated at 1000°C has four peaks in the case of the $^5D_0 \rightarrow {}^7F_1$ transition, more than the number allowed by 2J+1. This is further evidence that the ion has a new surrounding.

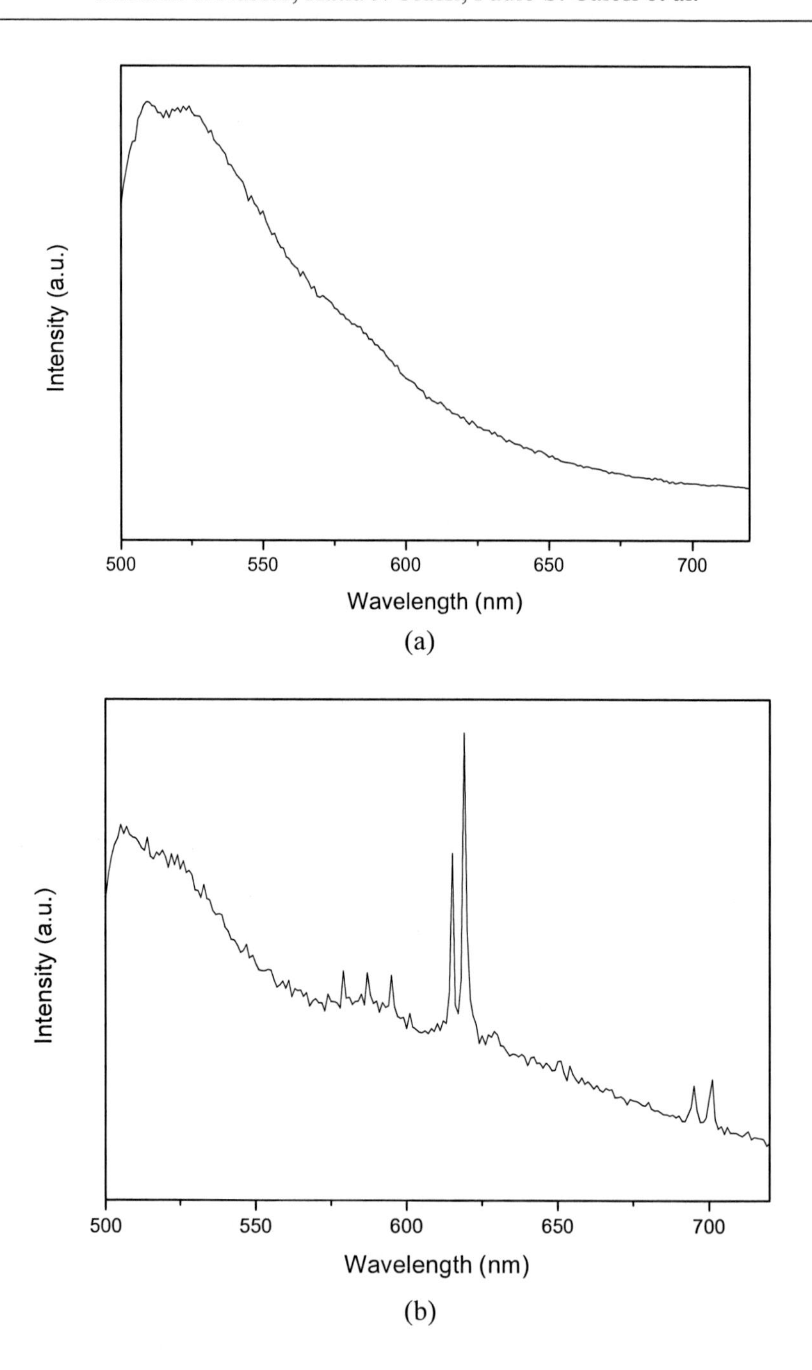

(a)

(b)

Figure continued

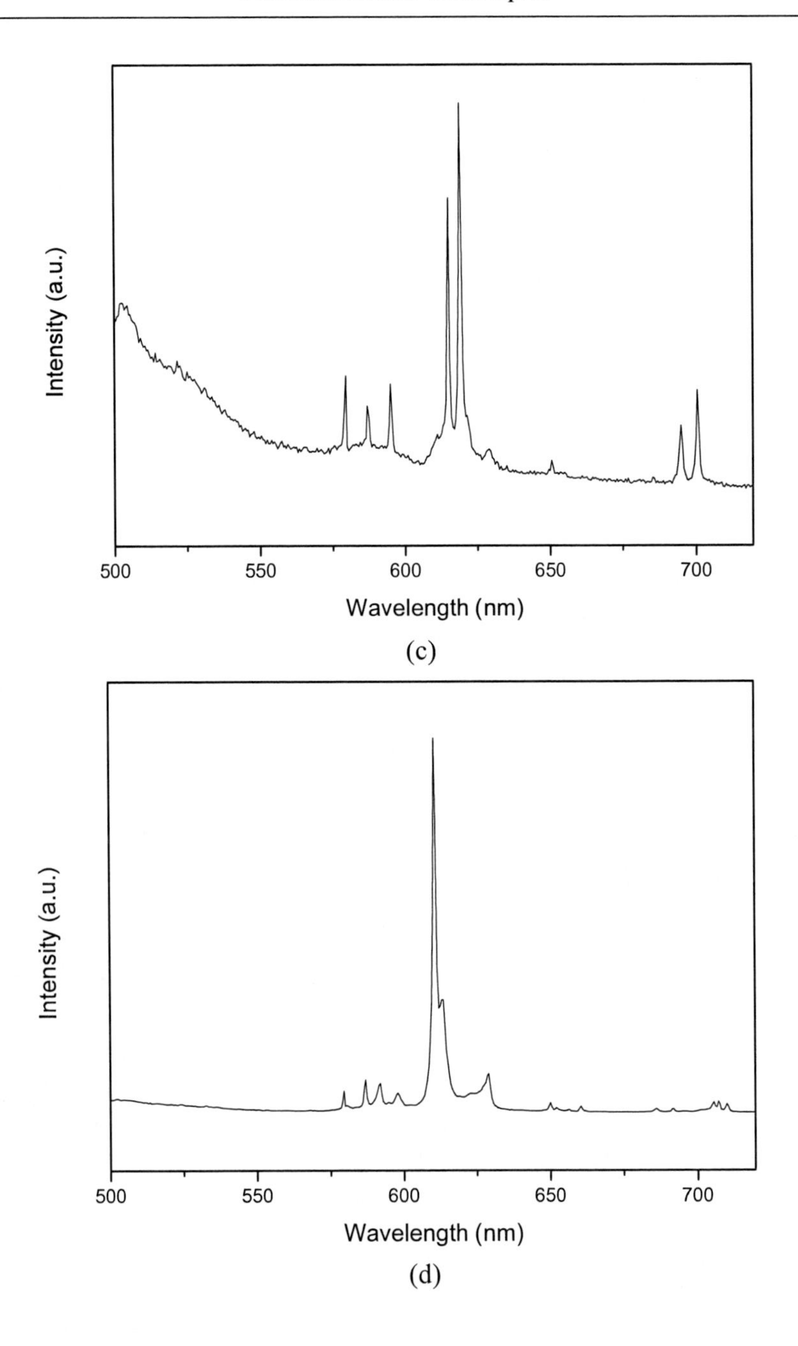

(c)

(d)

Figure continued

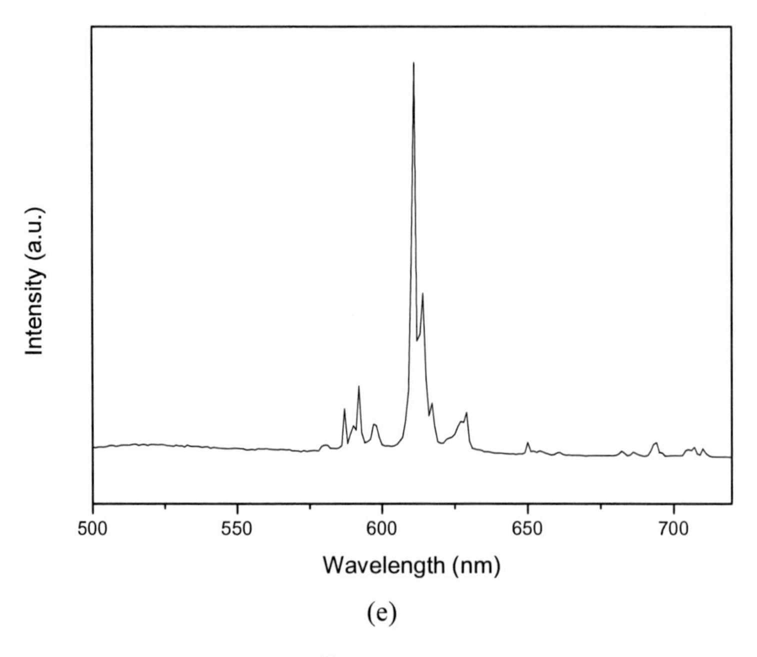

(e)

Figure 41. Emission spectra of the Eu^{3+} ion in the Gd(Eu)-Al-O samples treated at (a) 100, (b) 400, (c) 600, (d) 800, and (e) 1000°C for 4 hours, excited at 467 nm ($^5D_2$ level).

Because Gd^{3+} and Eu^{3+} are very similar in terms of ionic radius, the doped Eu^{3+} ion occupies the Gd^{3+} sites, thereby resulting in the hypersentitive red emission transition $^5D_0 \rightarrow {}^7F_2$ of Eu^{3+} , the most prominent group in the emission of this ion [99].

The luminescence quantum efficiency is determined by a balance between radiative and non-radiative processes, and can be estimated by solving a set of appropriate rate equations that involve the transitions and energy transfer rates, as well as the populations of the energy levels of both lanthanide ions [100].

The samples excited at CTB have one lifetime only, indicating that energy transfer occurs in only one kind of Eu^{3+} site. The biexponential decay behavior of the activator is frequently observed when the excitation energy is transferred to different Eu^{3+} sites from the donor. The samples excited at the $^5L_6$ and $^5D_2$ levels have two lifetimes, which is one more indication of the presence of different Eu^{3+} sites in the host, as demonstrated by the XRD and photoluminescence techniques.

In general, excitation within the *f*-manifold promotes high quantum efficiencies, whereas charge transfer excitation yields low efficiency. We observed one lifetime and higher quantum efficiency for the samples excited at CTB. Excitation at CTB reveals that energy transfer takes place in only one Eu^{3+} site, while the excitation in *f*-manifold ($^5L_6$ and $^5D_2$ levels) shows the existence of more than one Eu^{3+} site. Energy transfer to different types of site can reduce the quantum efficiency.

When we compared the luminescence properties of the Eu^{3+} ions incorporated into matrices with and without Ca^{2+} ion, the emission spectra was better defined in the case of the matrix without Ca^{2+} ion, indicating that Eu^{3+} occupies the Gd symmetry sites.

Another possibility for the preparation of phosphors by the non-hydrolytic sol-gel methodology is to change the gadolinium ion for yttrium, in order to obtain a crystalline phase with different symmetry sites for the Eu^{3+} ion.

The preparation of the gel containing yttrium was carried out in oven dried glassware. The material was synthesized via a modification of the method described by Acosta *et al.*[6]. Aluminum chloride ($AlCl_3$) in ethanol (EtOH) was reacted with yttrium chloride at different molar ratios, and. 1% of $EuCl_3$ in relation to the aluminum molar ratio was added as structural probe. The mixture remained in reflux at 110°C for 3 h, in 50 mL of previously distilled, dry DCM, under argon atmosphere. The condenser was placed in a thermostatic bath at –5°C. After 120 min, a solid material was obtained. After reflux, the mixture was cooled and aged overnight in the mother liquor at room temperature (RT), because precipitation continues upon aging in the mother liquor [101]. The solvent was then removed under vacuum. The obtained powders were submitted to several treatments, such as the use of different heat-treatment temperatures, namely 400, 800, 1100, and 1500°C, for 3 hours. Also, treatments using the same temperature but different periods of time were accomplished. The scheme 10 presents the structure of perovskite.

In another case, the mixture was homogenized by microwave radiation. To this end, the solid solution was placed on a SiC plate (2 x 1 x 0.5 cm) and submitted to microwave irradiation for 30 s or 2 min, in a conventional household equipment (Consul, model CMU31A, frequencies 2.45 GHz, power 1500 W). A temperature of 1000°C was reached.

The photoluminescence emission spectra (Figures 42 and 43) of the samples display a broad band between 250 and 300 nm, assigned to CTB transitions in the Eu^{3+}–O^{2-} bond lying in the band gap region of the Y_2O_3 host matrix [102, 4].

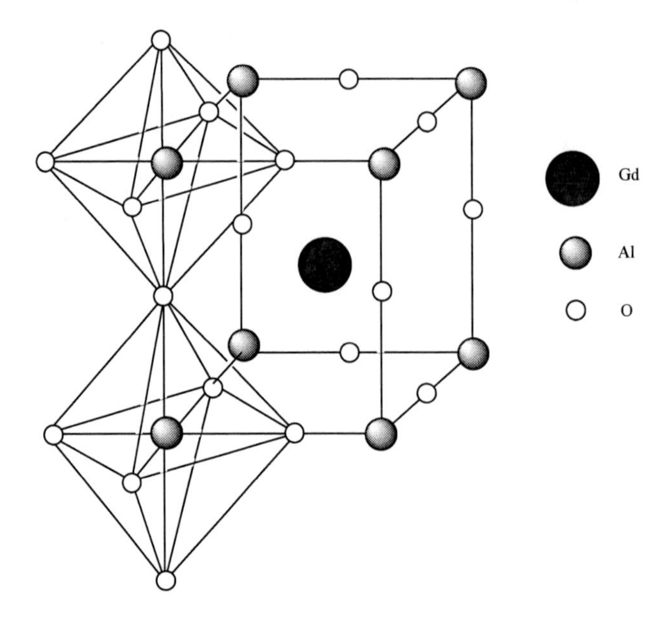

Scheme 10. Symmetry site in the peroviskite structure.

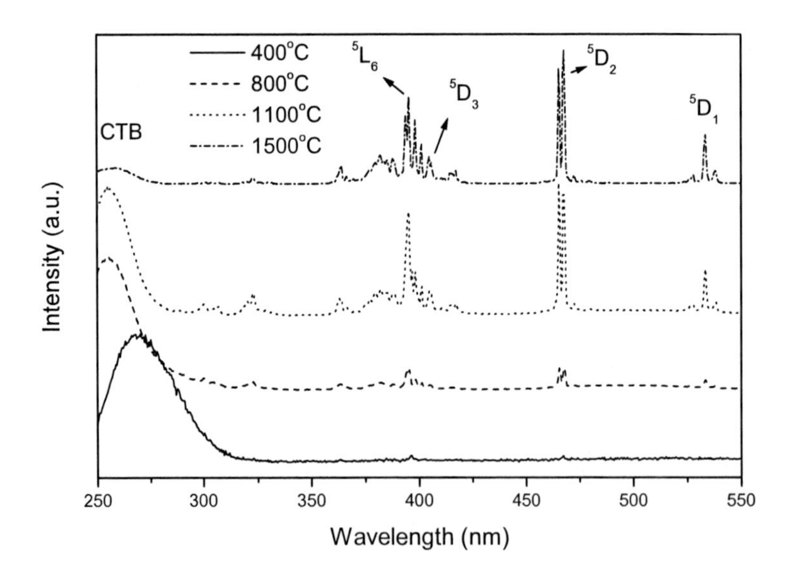

Figure 42. Excitation spectra of the sample containing Al/Y molar ratio of 1:1 heat-treated at different temperatures, $^{5}L_{6}$ (395 nm), $^{5}D_{3}$ (405 nm), $^{5}D_{2}$ (468 nm), and $^{5}D_{1}$ (533 nm).

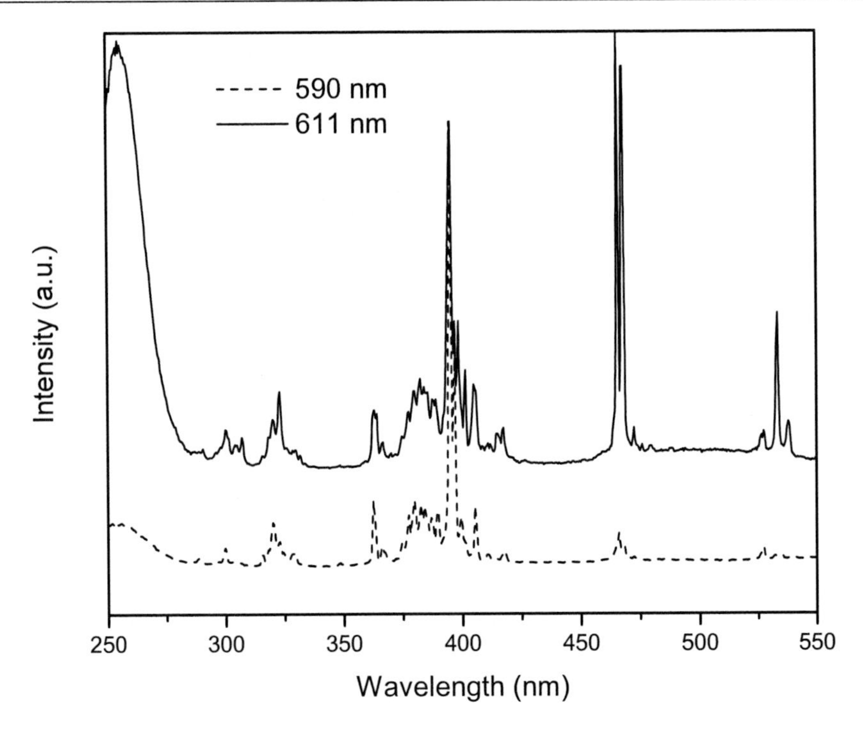

Figure 43. Excitation spectra of the sample containing Al/Y molar ratio 1:1 heat-treated at 1100°C, monitored at 590 ($^5D_0 \rightarrow {}^7F_1$) and 611 nm ($^5D_0 \rightarrow {}^7F_2$).

The sharp lines in the PLE spectra of the Eu^{3+}-doped aluminum-yttrium oxide samples with Al/Y molar ratio of 1:1, monitored at 620 and 611 nm ($^5D_0 \rightarrow {}^7F_2$, Figure 42, were assigned to transitions between the $^7F_0$ and the $^5L_6$, $^5D_{1-3}$ levels. As for the the PLE spectra of the Eu(III)-doped aluminum-yttrium oxide sample with Al/Y molar ratio of 1:1 and treated at 1100°C, monitored within the $^5D_0 \rightarrow {}^7F_1$ (590 nm) and $^5D_0 \rightarrow {}^7F_2$ (611 nm) transitions, they revealed that the $^5D_0 \rightarrow {}^7F_1$ transition is preferentially excited at 395 nm ($^5L_6$ levels), whereas the $^5D_0 \rightarrow {}^7F_2$ transition is favored by the excitation at 466 nm ($^5D_2$ levels).

Table 17 presents the maximum CTB for the samples investigated here. It is noteworthy that the CTB appears in the same spectral region for all the samples containing yttrium. However, in the case of the samples containing no yttrium, the band is located at different wavelengths, thereby confirming its charge transfer nature.

Table 17. Maximum CTB (λ_{max} in nm) for the samples with different yttrium contents and different heat-treatment temperatures

	Temperature (oC)				
Samples	100	400	800	1100	1500
	λmax (nm)				
Al:Y (1:0)	-	-	284	296	-
Al:Y (1:0.25)	-	291	261	261	264
Al:Y (1:0.50)	-	291	262	263	258
Al:Y (1:0.75)	-	292	259	257	260
Al:Y (1:1)	-	287	256	255	260

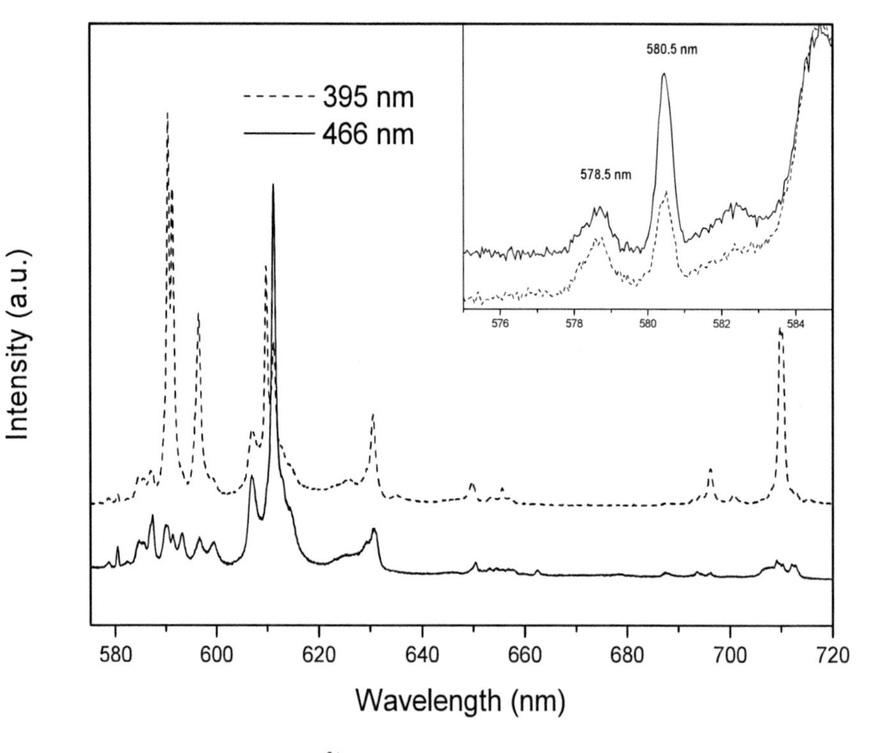

Figure 44. Emission spectra of Eu^{3+}-doped aluminum-yttrium (1:1) matrix treated at 1000°C recorded at different excitation wavelengths. Insert $^5D_0 \rightarrow {}^7F_0$ transition of Eu^{3+}.

The spectral properties of Eu^{3+} undergo major changes ongoing from the amorphous to the crystalline state. Indeed, small changes in site symmetries affect both the intensity and position of the emission bands. The emission spectra of the Eu^{3+}-doped aluminum-yttrium matrix (1:1) were recorded for several excitation wavelengths (395–466 nm), Figures 44 and 45, for the samples treated at 1100 and 1500°C, respectively. The corresponding inserts present the details of the $^5D_0 \rightarrow {}^7F_0$ regions, and they indicate the presence of more than one local Eu^{3+} environment. This is inferred from the presence of two distinct $^5D_0 \rightarrow {}^7F_0$ transitions located at 578.5 and 580.5 nm for the samples treated at 800 and 1100°C and recorded at an excitation wavelength of 395 nm, at these respective temperatures. When a wavelength excitation of 466 nm was employed, three distinct bands were detected in this spectral region at ca. 578.5, 580.5, and 582.3 nm. The sample treated at 1500°C presented three distinct bands, too.

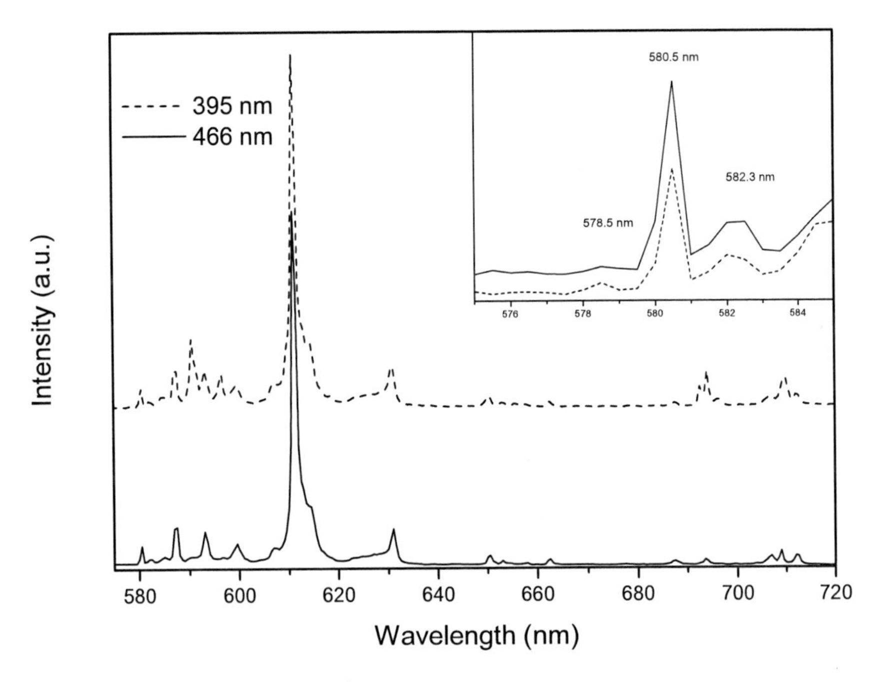

Figure 45. Emission spectra of Eu^{3+}-doped aluminum-yttrium (1:1) matrix treated at 1500°C recorded at different excitation wavelengths. Insert $^5D_0 \rightarrow {}^7F_0$ transition of Eu^{3+}.

Figure 46 shows the band corresponding to the $^5D_0 \rightarrow {}^7F_1$ transition for the sample containing Al/Y at a 1:1 molar ratio.

These spectra show that the $^5D_0 \rightarrow {}^7F_1$ transition is preferentially excited at 395 nm ($^5L_6$ levels), whereas the $^5D_0 \rightarrow {}^7F_2$ transition is favored by the excitation of the level 466 nm ($^5D_2$ levels). In fact, Rainho *et al.* observed the same behavior for Eu^{3+}-doped synthetic narsarsukite [103]. The six bands in the $^5D_0 \rightarrow {}^7F_1$ region confirm the existence of more than one Eu^{3+} ion site. This transition (J = 1) presents a maximum number of emission bands equal to 3. The luminescence of Eu^{3+} in the aluminum-yttrium matrix treated at 1100°C occurs mainly via the $^5D_0 \rightarrow {}^7F_1$ transition, indicating that the Eu^{3+} ions lie in centrosymmetrical sites. The X-ray diffraction revealed that the predominant phase at 1100°C is the cubic $Al_5Y_3O_{12}$ (YAG) [104]. At 1500°C, the predominant phase is monoclinic $Al_2Y_4O_9$ (YAM). The scheme 11 presents the structure of YAG.

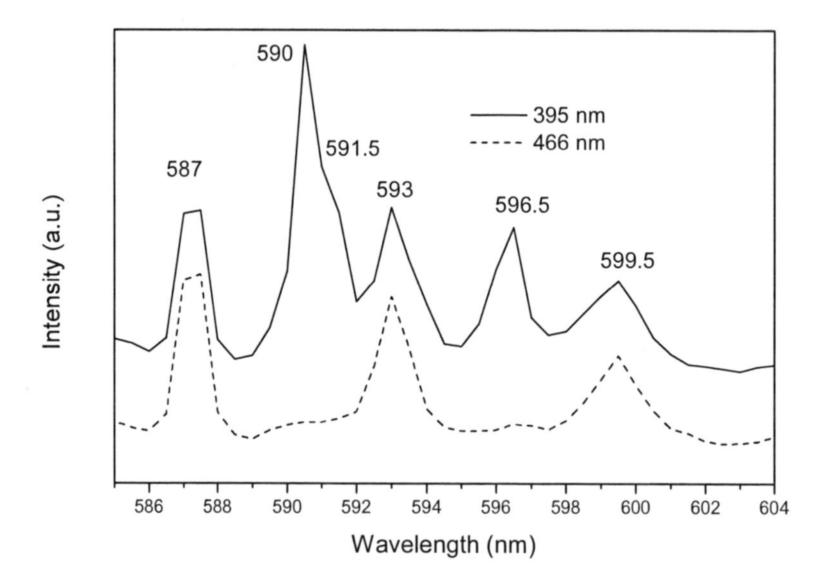

Figure 46. Emission spectra of the $^5D_0 \rightarrow {}^7F_1$ transition of Eu^{3+} doped into aluminum-yttrium (1:1) matrix, recorded at different excitation wavelengths for the samples treated at 1500°C.

Lifetime measurements were performed for different excitation wavelengths, namely 395 and 466 nm, while emission was monitored at the peaks of 590 and 611 nm, respectively. The lifetime was high for the phase $Al_5Y_3O_{12}$ excited at 395 nm and recorded at 590 nm, around 3.80 ms. As for the $Al_2Y_4O_9$ phase, the lifetime upon excitation at 466 nm was 1.50 ms for the emission recorded at 611 nm. There was a decrease in the lifetime of the $^5D_0$

$\rightarrow {}^7F_2$ transition due to temperature. This fact can be explained by sample synterization.

In conclusion, the substitution of gadolinium with yttrium ions leads to a new symmetry site. A site with an inversion center and another site without inversion center can be seen in the emission spectra of the Eu^{3+} ion doped into the aluminum-yttrium matrix (Figure 44).

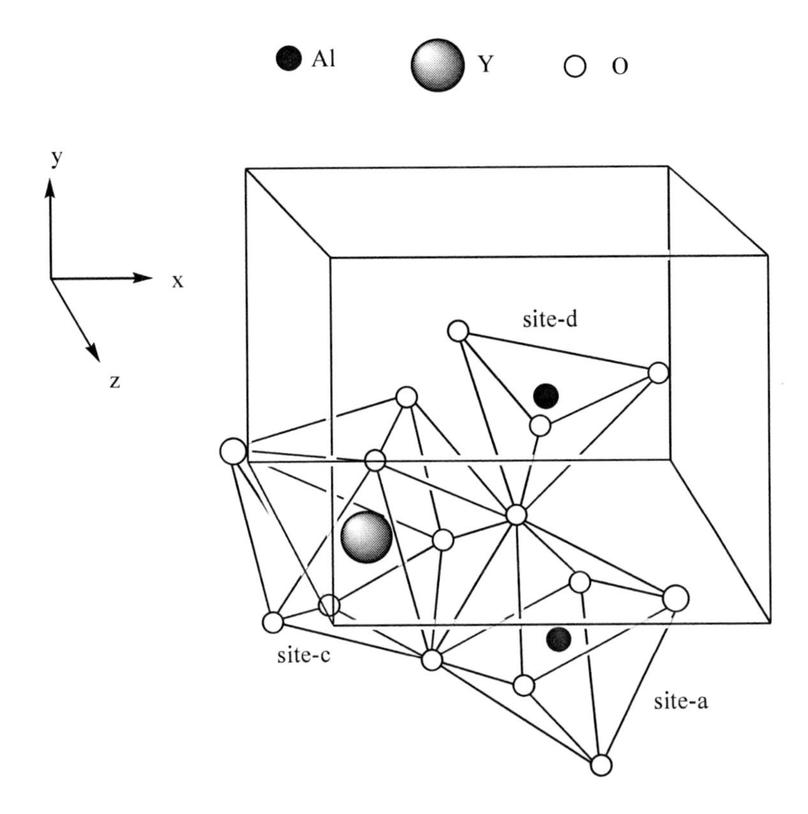

Scheme 11. Symmetry site in the YAG structure.

Chapter 4

CONCLUSION

In this work we have shown the different emission spectra of the Eu^{3+} ions in several matrices obtained by the sol-gel methodoly, which have furnished information about the surroundings of this ion. The environment around Eu^{3+} depends on the methodology employed for matrix preparation as well as on matrix composition. The matrices obtained by the hydrolytic sol-gel route contain silicon as the main component, and the emission spectra of the Eu^{3+} ion in these matrices evidenced formation of an amorphous phase. As for the non-hydrolytic sol-gel methodology, production of crystalline samples is favored, although in several cases a phase mixture is detected. In some cases, the same sample presents symmetry sites without or with inversion centers, as detected in the emission spectra of the Eu^{3+} ion doped in such matrices.

ACKNOWLEDGMENTS

The authors acknowledge FAPESP, CNPq, and CAPES (Brazilian research funding agencies) for financial support.

REFERENCES

[1] Martins, TS; Isolani, PC. Quím. *Nova*, 2005, 28(1), 111-117.

[2] Sastry, VS; Bunzli, JC; Perumareddi, JR; Rao, VR; Rayudu, GVS. In. Modern Aspects of Rare Earths and their Complexes" *Hardbound*, 1006 pages, publication date: DEC-2003.

[3] Horrocks, Jr., W. de, W; Albin, M. *Prog. Inorg. Chem.*, 1984, 31(1), 1-103.

[4] Blasse, G; Grabmaier, BC. *Luminescent Materials*, Springer-Verlag Berlin Heidelberg 1994.

[5] Hench, LL; West, JK. *Chemical Reviews*, 1990, 90(1), 33-72.

[6] Acosta, S; Corriu, RJP; Leclerq, D; Lefervre, P; Mutin, PH; Vioux, AJ. *Non-Cryst. Solids*, 1994, 170, 234.

[7] Nassar, EJ; Pereira, PFS; Ciuffi, KJ; Calefi, PS. "Recent Development of luminescent materials prepared by the sol-gel process", In: Photoluminescence Research Progress, chapter 10, Editors: K; Harry, Wright, V; Grace, *Edwards*, 2008 Nova Science Publishers, Inc.

[8] Wright, JD; Sommerdijk, NAJ. In: *Sol-Gel Materials Chemistry and Applications*, Taylor and Francis Books 2003.

[9] Brinker, CJ; Scherer, GW. Sol-Gel Science, *The Phys Chem. Sol-Gel Processing*, Academic Press, San Diego, 1990.

[10] Nassar, EJ; Avila, LR; Pereira, PFS; Nassor, ECO; Cestari, A; Ciuffi, KJ; Calefi, PS. *Quím*. Nova, 2007, 30(7), 1567-1572.

[11] Avila, LR; Nassor, EC. De, O; Pereira, PFS; Cestari, A; Ciuffi, KJ; Calefi, PS; Nassar, EJ. *J. Non-Cryst. Solids*, 2008, 354, 4806-4810.

[12] Nassar, EJ; Nassor, E. C. de O; Avila, LR; Pereia, PFS; Cestari, A; Luz, M; Ciuffi, KJ; Calefi, PS. *J. Sol-Gel Scie. Techn.*, 2007, 43, 21-26.

[13] Nassor, E C. de O; Avila, LR; Pereira, PFS; Ciuffi, KJ; Calefi, PS; Nassar, EJ. submitted, *Materials Research*, 2010.

[14] Nassar, EJ; Ciuffi, KJ; Ribeiro, SJL; Messaddeq, Y. *Materials Research*, 2003, 6(4), 557-562.

[15] Bandeira, LC; Avila, LR; Cestari, A; Calefi, PS; Ciuffi, KJ; Nassar, EJ; Salvado, IM; Fernandes, MHFV. submitted, *Quím*. Nova 2009.

[16] Bandeira, LC; Ciuffi, KJ; Calefi, PS; Nassar, EJ. submitted *Materials Chemistry and Physics*, 2009.

[17] Rocha, LA; Avila, LR; Caetano, BL; Molina, EF; Sacco, HC; Ciuffi, KJ; Calefi, PS; Nassar, EJ. *Materials Research*, 2005, 85(2-3), 361-362.

[18] Rocha, LA; Ciuffi, KJ; Sacco, HC; Nassar, EJ. *Mater. Chem. Phys.*, 2004, 85(2-3), 245-250.

[19] Rocha, LA; Molina, EF; Ciuffi, KJ; Calefi, PS; Nassar, EJ. *Mater. Chem. Phys.*, 2007, 101(1), 238-241.

[20] Nassar, EJ; Ciuffi, KJ; Gonçalves, RR; Messaddeq, Y; Ribeiro, SJ. L. *Quím*. Nova 2003, 26(5), 674-677.

[21] de Souza, FJ; de Lima, GPA; Pereira, PFS; Avila, LR; Ciuffi, KJ; Nassar, EJ; Calefi, PS. *Materials Research*, 2010 in press.

[22] Cestari, A; Avila, LR; Nassor, E. C. de O; Pereira, PFS; Calefi, PS; Ciuffi, KJ; Nakagaki, S; Gomes, ACP; Nassar, EJ. *Materials Research*, 2009, 12(2), 139-143.

[23] Cestari, A; Bandeira, LC; Calefi, PS; Nassar, EJ; Ciuffi, KJ. *J. Alloys Compd*. 2009, 472, 299-306.

[24] Matos, MG; Pereira, PFS; Calefi, PS; Ciuffi, KJ; Nassar, EJ. *J. Lumin*, 2009, 129, 1120-1124.

[25] Matos, MG; Calefi, PS; Ciuffi, KJ; Nassar, EJ. submitted, *Mater. Chem. Phys.*, 2010.

[26] Nassar, EJ; Avila, LR; Pereira, PFS; Mello, C; de Lima, OJ; Ciuffi, KJ; Carlos, LD. J. *Lumin*, 2005, 111, 159-166.

[27] Nassar, EJ; Avila, LR; Pereira, PFS; de Lima, OJ; Rocha, LA; Mello, C; Ciuffi, KJ. *Quím*. Nova 2005, 28(2), 238-243.

[28] Pereira, PFS; Caiut, JMA; Ribeiro, SJL; Messaddeq, Y; Ciuffi, KJ; Rocha, LA; Molina, EF; Nassar, EJ. *J. Lumin*, 2007, 126, 378-382.

[29] Nassar, EJ; Pereira, PFS; Nassor, E. C. de O; Avila, LR; Ciuffi, KJ; Calefi, PS. J. Mater. *Scie*., 2007, 42, 2244-2249.

[30] Ciuffi, KJ; de Lima, OJ; Sacco, HC; Nassar, EJ. *J. Non-Cryst. Solids*, 2002, 304, 126- 133.

[31] Levy, D; Reisfeld, R; Avnir, D. *Chem. Phys. Lett*, 1984, 109, 593.

[32] Zhou, L; Choy, WCH; Shi, J; Gong, M; Liang, H; Yuk, TI; *J. Solid State Chem.*, 2005, 178, 3004.

[33] Forest, H; Ban, G. *J. Electrochem. Soc.: Solid State Science*, 1969, 116(4) 474.

[34] Hazenkamp, MF; Van der Veen, AMH; Feiken, W; Blasse, G. J. *Chem. Soc. Faraday Trans*. 1992, 88(1), 141.

[35] Rice, DK; Deshase, LG; *Phys. Rev. B*, 1969, 186.

[36] Reisfeld, R. J. *Eletrochem. Soc*., 1984, 131, 1360.

[37] Richardson, FS. *Chem. Rev*., 1982, 82, 541.

[38] Reisfeld, R. *Struct*. Bond. 1979, 30, 65.

[39] Serra, OA; Campos, RM. *Quím*. Nova 1991, 14(3), 159-161.

[40] Legendziewicz, J; Guzik, M; Cybin`ska, J. *Optical Materials*, 2009, 31, 567-574.

[41] Matraszek, A; Macalik, L; Szczygiel, Godlewska, P; Solarz, P; Hanuza, J; *J. Alloys Compds*, 2008, 451, 254-257.

[42] Caiut, JMA; Bazin, L; Mauricot, R; Dexpert, H; Ribeiro, SJL; Dexpert-Ghys, J. *J. Non-Cryst. Solids*, 2008, 354, 4860-4864.

[43] Carlos, LD; Ferreira, RAS; Bermudez, VZ; Ribeiro, SJ. L. *Adv. Funct. Mater*, 2001, 11, 111.

[44] Blasse, G. Adv. Inorg. *Chem*., 1990, 35, 319.

[45] Caiut, JMA; Rocha, LA; Sigoli, FA; Messaddeq, Y; Dexpert-Ghys, J; Ribeiro, SJL. J. Non-Cryst. *Solids*, 2008, 354, 4795-4799.

[46] Wojtach, K; Cholewa-Kowalska, K; Laczka, M; Olejinizak, Z. *Optical Materials*, 2005, (27), 1495.

[47] Godoi, RHM; Fernandes, L; Jafelicci Jr, M; Marques, RC; Varanda, LC; Davolos, MR. *J. Non-Cryst. Solids*, 1999, 247, 141-145.

[48] Vinod, PM; Bahnemann, Rajamonhana, RB; Vijayamohanan, K. *J. Phys. Chem. B*, 2003, 107, 11583.

[49] Anagnostopoulos, T; Eliades, G; Palaghias, G. *Dental Materials*, 1993, 9, 182.

[50] Skoog, DA. In: Princípios de Análise Instrumental; trad. Ignez Caracelli. – 5.ed. – Porto Alegre: *Bookman*, 2002.

[51] Chiang, CH; Ishida, H; Koenig, JL. *J. Coll. Inter. Scie*., 1980, 74(2), 396.

[52] Ribeiro, SJL; Hiratsuka, RS; Massabini, AMG; Davolos, MR; Santilli, CV; Pulcinelli, SH. *J. Non-Cryst. Solids*, 1992, 147/148, 162.

[53] Habemchuss, A; Spedding, FH. *J. Chem. Phys*., 1980, 73, 442.

[54] Serra, OA; Nassar, EJ; Zapparolli, G; Rosa, ILV. *J. Alloys Comp*., 1994, 207/208, 454.

[55] Reisfeld, R. *Structure Bonding*, 1973, 13, 53.
[56] Kokubo T. *Biomaterials*, 1991, 12(1), 55-63.
[57] Villacampa, AI; Ruiz, JMG. J. *Crystal Growth*, 2000, 11, 111-115.
[58] Kake, MA; Joshi, CP; Moharil, SV; Muthal, PL; Dhopte, SM. *J. Lumin*, 2008, 128, 1225.
[59] Ravichandran, D; Roy, R; Chaklowshoi, G; Hunt, C; White, WB; Erdei, S. J. *Lumin*, 1997, 71, 291.
[60] Bao, A; Tao, C; Yong, H. J. *Lumin*, 2007, 126, 859.
[61] Ternane, R; Trabelsi-Ayedi, M; Kbir-Ariguid, N; Piriou, B. J. *Lumin*, 1999, 81, 165.
[62] Jagannathan, R; Kottaisamy, M. J. *Phys*., Condens Mater 1995, 7, 8453 - 8466.
[63] Piriou, B; Fahmi, D; Dexpert-Ghys, J; Taitai, A; Lacout, JL. J. *Lumin*, 1987, 39, 97-103.
[64] Mehanaouni, M; Panczer, G; Ternane, R; Trabelsi-Ayedi, M; Boulon, G. *Optical Materials*, 2008, 30, 672.
[65] Yang, Y; Ren, Z; Tao, Y; Cui, Y; Yang, H. *Current Applied Physics*, 2009, 9, 618.
[66] Lin, H; Yang, D; Liu, G; Ma, T; Zhai, B; An, Q; Yu, J; Wang, X; Liu, X; Pun J. *Lumin*, 2005, 113, 121.
[67] Krebs, JK; Brownstein, JM. *J. Lumin*, 2007, 124, 257-259.
[68] Krebs, JK; Brownstein, JM; Gibides, JT. *J. Lumin*, 2008, 128, 780-782.
[69] Park, E; Condrate, RA; Lee, D; Kociba, J; Gallagher, PK. *J. Mater. Sci.: Materials in Medicine*, 2002, 13, 211.
[70] de Sá, GF; de Azevedo, WM; Gomes, ASL. *J. Chem. Res*., 1994, 234 .
[71] Sager, WF; Filipescu, N; Serafin, FA. *J. Phys. Chem*., 1965, 69, 1092.
[72] Jorgensen, CK; Reisfeld, RJ. Less. *Common Met*, 1983, 83, 107.
[73] Reisfeld, R. *Structure Bonding*, Springer, New York, 1975.
[74] Sibu, CP; Kumar, SR; Mukundan, P; Warrier, KGK. *Chem. Mater*., 2002, 14, 2876.
[75] Negishi, N; Takeuchi, K. Mater. *Letters*, 1999, 38, 150.
[76] Souza, LA; Messaddeq, Y; Ribeiro, SJL; Federicci, C; Lanciotti Jr., F; Pzani, PS. *Quím*. Nova 2002, 25(6B), 1067.
[77] Avnir, D. Acc. *Chem. Rev*., 1995, 28, 328.
[78] Rosa, ILV; Serra, OA; Nassar, EJ. *J. Lumin.* 1997, 72-74, 532.
[79] Reisfeld, R; Zelner, M; Patra, A. J. *Alloys Compd*., 2000, 300-301, 147.
[80] de Oliveira, E; Neri, CR; Serra, OA; Prado, AGS. *Chem. Mater*. 2007, 19, 5437-5442.

[81] Escribano, P; Júlian-Lopez, B; Planelles-Aragó, J; Cordoncillo, E; Vianna, B; Sanchez, C. J. *Mater. Chem.*, 2008, 18, 23.

[82] Gardolinsky, JE; Martins Filho, HP; Wypyck, F. *Quim.* Nova. 2003, 26, 30.

[83] Gardolinsky, JE; Ramos, LP; Souza, GP; Wypych, F. J. Coll. *Int. Sci.*, 2000, 221, 284.

[84] Tunney, JJ; Detellier, C; Can. J. *Chem.*, 1997, 75, 1766.

[85] de Faria, E. H; de Lima, OJ; Ciuffi, KJ; Nassar, EJ; Vicente, MA; Trujillano, R; Calefi, PS. *J. Coll. Inter. Sci.*, 2009, 335(2), 210-215.

[86] Tunney, JJ; Detellier, C. *Mater.*, 1993, 5, 747.

[87] Brandt, KB; Elboki, TA; Detellier, C. *J. Mater. Chem.*, 2003, 13, 2566.

[88] Rinaldi, R; Schuchardt, U. J. *Catal.* 2004, 227, 109.

[89] Reinhard, C; Gudel, H. U. *Inorg. Chem.*, 2002, 41, 1048.

[90] Binnemans, K. Rare Earth Beta- Diketonates. Chapeter 225. Handbook on the Physics and Chemistry of Rare Earth. *Elservier,* 2005, 35, 107.

[91] Zolin, VF; Puntus, LN; Tsaryuk, VI; Kudryashova, VA; Legendziewicz, J; Gawryszewska, P; Szostak, R. J. *Alloys Compds*, 2004, 380, 279.

[92] Schneider, RF. Titration of Hydrogen Peroxide Solutions, apostila disponível no sítio: www.ic.sunysb.edu/Class/che133/susb/susb035.pdf (último acesso em 5 de julho de 2008).

[93] Nassar, EJ; Serra, O. A. *Quim.* Nova. 2000, 1(23), 16.

[94] Griffin, SG; Hill, RG. *Biomaterials*, 1999, 20, 1579-1586.

[95] Phillips, RW. "Skinner Materiais Dentários", Editora Guanabara Koogan, 9a Edição, Rio de Janeiro, RJ, 1993.

[96] Bunzli, JCG; Choppin, GR. Lanthanide Probes in Life, *Chemical and Earth Sciences*, Elsevier, 1989.

[97] Krebs, JK; Brownstein, JM; Department of Physics and Astronomy, Franklin & Marshall College, *Lancaster*, 2006, 17604-3003.

[98] Gorman, CM; Hill, R. G. *Dental Materials*, 2003, 19, 320–326.

[99] Wu J; Yan, BJ. *Alloys Compds*, 2008, 432, 293-297.

[100] de Sá, GF; Malta, OL; Donegá, CM; Simas, AM; Longo, RL; Santa-Cruz, PA; da Silva, Jr. E. F. *Coordiation Chem. Rev.*, 2000, 196, 165-195.

[101] de Lima, OJ; Aguirre, DP; de Oliveira, DC; da Silva, MA; Mello, C; Leite, CAP; Sacco, HC; Ciuffi, KJ. *J. Mat. Chem.*, 2001, 11, 2476.

[102] Dorota, AP; Wozniak, K; Frukacz, Z; Barr, TL; Fiorentino, D; Seal, S. *J. Phys. Chem. B*, 1999, 103, 1454.

[103] Rainho, JP; Lin, Z; Rocha, J; Carlos, LD. J. Sol-Gel Sci. *Technology*, 2003, 26, 1005.

[104] Ruan, SK; Zhou, JG; Zhong, AM; Duan, JF; Yang, XB; Su, MZ. J. *Alloys Compds*., 1998, 275-277, 72.

INDEX

F

G

H

I

L

M

N

O

P

R

S

T

U

V

W

X

Y